CERAMIC SCIENCE FOR THE POTTER

SECOND EDITION

CERAMIC SCIENCE FOR THE POTTER

W.G. LAWRENCE & R.R. WEST

CHILTON BOOK COMPANY · RADNOR, PENNSYLVANIA

Dedicated to Dr. John F. McMahon, Dean Emeritus of the New York State College of Ceramics, King of the Leprechauns, friend, counselor and inspiration to ceramists everywhere.

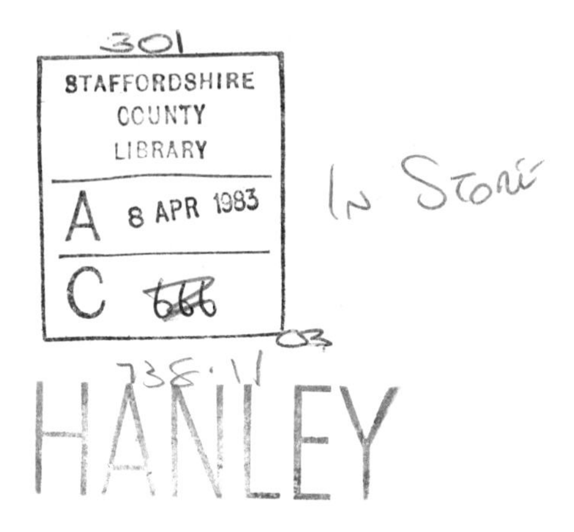

Published in Radnor, Pennsylvania 19089, by Chilton Book Company
and simultaneously in Canada by VNR Publishers,
1410 Birchmount Road, Scarborough, Ontario M1P 2E7
Designed by Jean Callan King/Visuality
Manufactured in the United States of America

Library of Congress Cataloging in Publication Data

Lawrence, W. G. (Willis Grant), 1916–
Ceramic science for the potter.
Includes bibliographies and index.
1. Ceramic materials. 2. Pottery.
I. West, R. R. (Richard Rudolph), 1919–
II. Title.
TP810.5.L38 1982 666 81-70913
ISBN 0-8019-7155-1 AACR2

1 2 3 4 5 6 7 8 9 0 1 0 9 8 7 6 5 4 3 2

CONTENTS

PREFACE

There are many reasons for preparing a revision of a book. Usually the mere passage of time results in new, useful information; the use of a book in the classroom pinpoints errors and useless or unimportant subject matter and always there is the need for expansion of the discussion of certain areas.

The first edition of *Ceramic Science for the Potter* was prepared primarily for the ceramic art potter even though the term *potter* should certainly include the large group of ceramic engineers who work with clay bodies in the whiteware industry. This industry includes dinnerware, sanitary ware and electrical porcelain manufacturers. The composition of whiteware bodies is very similar to that of the porcelains used by artists, and many common problems exist.

In expanding the subject matter to include the field of whiteware, the experience and expertise of Richard West, Professor Emeritus of Ceramic Engineering at the New York State College of Ceramics at Alfred University, has been most helpful. He has consulted, researched and taught in the field of ceramic whiteware for the past 35 years. His scientific contributions include numerous monographs and over 30 technical papers.

His application of computer analysis to the problems of thermal expansion of crystalline bodies and glaze fit is most significant. To my knowledge, his calculations are the first to predict thermal expansions in complex multicrystalline ceramic systems. He also uses computer analysis to predict such glaze properties as fusion temperature and thermal expansion. The story of the glaze fit problem involving the Blue Hippo and its solution through the use of computer techniques is most appropriate and interesting. Professor West's industrial consulting experience is evident in that his contributions are directly related to manufacturing problems, yet many of these are common to both the artist and engineer.

In addition to the broadening of the subject matter in the field of whiteware, some of the more arcane sections on atomic structure, structure of silicates and phase diagrams have been greatly reduced or eliminated.

CERAMIC SCIENCE FOR THE POTTER

CHAPTER 1

AN INTRODUCTION TO CERAMIC MATERIALS

1
An Introduction to Ceramic Materials

The ceramic artist, potter or designer makes use of many materials to create useful and appealing objects. Often, the potter has little background knowledge as to how these materials may behave. Unexpected or disappointing results may ultimately be corrected on a trial-and-error basis but usually at a great cost in time and effort. For those who have made such mistakes and who wish to understand their materials more thoroughly, this book will be useful.

Some say that a little understanding may be worse than none at all, but it is our experience that people with no scientific background still exhibit the intellectual curiosity and desire to know why things happen. Such knowledge enables the potter to prevent mistakes or to correct them quickly once the cause is recognized. We hope this book will help potters to more quickly arrive at the cause and solutions.

Pottery-making may appear to be a rather simple, uncomplicated series of steps starting with the mixing of a suitable clay with water, forming a shape, drying, firing and perhaps glazing. These are the basic steps of the process, but as the potter well knows, there are a number of variables that must be taken into account within each step, as must the interdependency among steps. The disconcerting thing about this interdependency is that a mistake in one of the preliminary or initial steps, such as forming, may not become evident until the final product has been in use for some time.

The materials the potter uses are far from simple. Although the basic starting material consists of clay plus an appropriate amount of water to produce the desired plastic properties, many other ingredients are added to impart desirable characteristics, for example, feldspars or other glass-forming materials, flint and calcined fireclay-grog. For the most part, the materials used in ceramic bodies are crystalline and remain so during the firing operation. Some glass is, of course, formed in firing, which sticks the crystalline materials together and provides the fired strength of the body. The properties exhibited in the final product are determined largely by the amount and types of crystals present and the relative amounts of crystalline and glassy phases present. Therefore, body additions should be based on a knowledge of the function of each ingredient added and its behavior with or reaction to the other body ingredients. The potter needs to know something about crystals, their structure and their thermal and physical properties.

The potter is also involved in making glazes that must properly fit the bodies he produces. Glazes, with few exceptions, are glasses. Glassy materials are entirely different from crystalline materials in that they are amorphous solids which have no repeating structure or orderly atomic arrangement. The potter should know something about the properties and characteristics of glasses, glass formation, viscosity, softening point and thermal expansion.

To provide a basis for an understanding of the differences between crystalline and glassy materials, the following brief and elementary descriptions are provided.

Crystalline Materials

The name crystal, from the Greek word *krystallos,* means clear ice, and was first applied to the transparent quartz crystals found in the Swiss Alps. It was believed that these "stones" were formed from water by intense cold. The name was later applied to all solids that had flat faces symmetrically arranged.

The essential difference between the liquid and solid or crystalline states is the difference in the relative mobility of the molecules or atoms. In liquids, the molecules are mobile, whereas in solids their attraction for each other holds them in apparently rigid positions and their movement is restricted. However, when some form of energy, such as heat, is applied, movement increases and produces such phenomena as solid state reactions, crystal inversions, decomposition and melting.

Well-formed crystals are relatively rare in nature. Only recently have techniques been perfected resulting in synthetic crystals. Because of this, crystals were formerly considered to be a rather mysterious exception to the more common form of solids. The microscope, invented late in the 16th century, made it evident that most materials are indeed crystalline.

The symmetrical characteristics as seen directly or with the aid of a microscope persist to much smaller dimensions in atoms and molecules themselves. The basic feature of a crystal is the regularity of the atomic arrangement. There is a basic unit in all crystal structures called the unit cell, which, when repeated in space, results in the entire crystal structure. Therefore, a crystal structure is a repetitive arrangement of atoms or molecules.

The manner in which atoms or ions arrange themselves to form crystals is determined mainly by their size and electrical charge, either positive or negative. In ceramic materials, it is the attraction between dissimilarly charged ions that holds them together in a solid form. Most ceramic materials are oxides composed of an oxygen ion having a negative charge, O^{-}, and some other positively charged ion, such as silicon, Si^{+4}. For example, in silicate structures the silicon-oxygen tetrahedron is the basic building block. The silicon ion and oxygen ions always form this tetrahedral structure, which is illustrated in Figure 1–1.

These silicon-oxygen tetrahedra can be connected in various patterns resulting in single chains, double chains and sheet-type structures, as shown in Table 1–1.[1]

The structure of clay minerals belongs to the silicate structure known as the sheet-type. Figures 1–2 and 1–3 show the structure of kaolinite and montmorillonite (bentonite).[1] This sheet-type structure of clay minerals is responsible for many of the problems posed by clays in forming, drying and firing. Crystal models and drawings may

Fig. 1–1 The silicon-oxygen tetrahedron, exploded perspective (left) and cutaway view (right).

help one to understand the atomic structure of clay particles, but they often seem quite abstract when one is dealing with clay particles themselves, which are too small to be seen in their surroundings, even with an electron microscope. Mica may be used as a model in assessing some of the characteristics of clays. Muscovite mica is almost identical in crystal structure to the clay mineral illite, and, of course, some of the chlorite minerals occur as finely divided clay minerals and/or as large fragments.

A large piece of muscovite mica may be held in the hand and examined in detail. The thin plates disclose a well-ordered sheet structure with ionic binding strong enough to resist tearing. Thinner and thinner sheets may be separated either with a knife or the fingernail because the bonding between sheets is of the much weaker

TABLE 1–1
SILICATE STRUCTURES

Name	No. of Shared Oxygens	Mineral Example	Silicon-Oxygen Arrangement
Orthosilicate	0	Forsterite Mg_2SiO_4	
Pyrosilicate	1	Melilite $Ca_2MgSi_2O_7$	
Metasilicate	2	Enstatite $MgSiO_3$	
Amphibole	2 & 3	Tremolite $Ca_2Mg_5Si_8O_{22}(OH)_2$	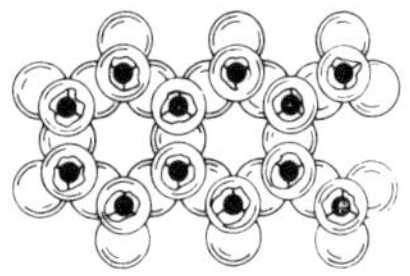
Disilicate	3	Kaolinite $Al_2Si_2O_5(OH)_4$	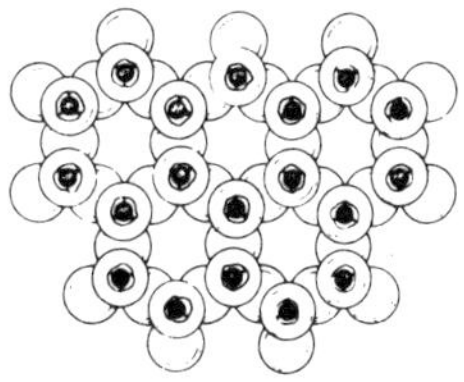
Three Dimensional	4	Quartz SiO_2	Not illustrated

Van der Waal type. Even kaolin minerals are similar to mica because kaolinite occurs in stacks that may be separated by mechanical forces.[2] This information has allowed kaolin suppliers to process the undesired kaolin stacks by delamination and provide the thin plates desired as a premium product for the paper industry.[3] These plates are also much whiter because, during the delamination process, a black deposit of the amorphous mineral leucoxene is removed. A waste product, kaolin stacks, has thus been converted into a premium product, kaolin plates.

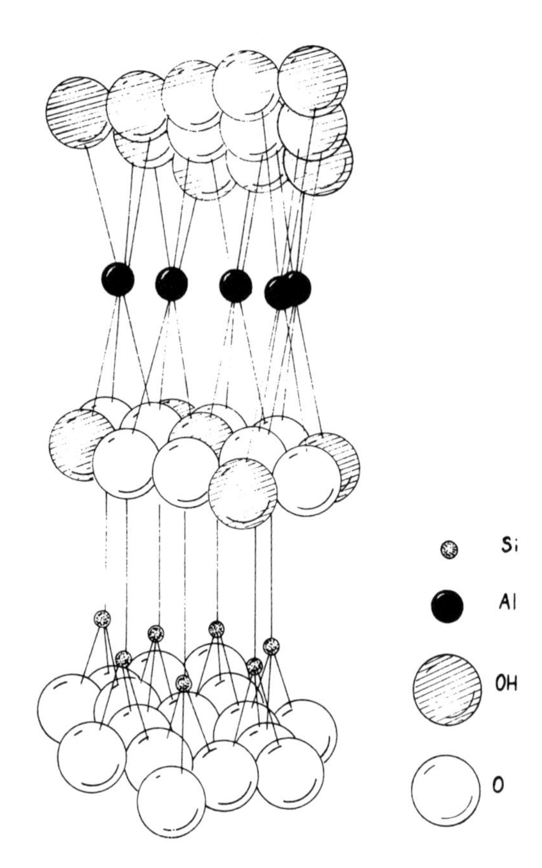

Fig. 1–2 Structure of kaolinite, $Al_2Si_2O_5(OH)_4$ illustrating the combination of tetrahedral and octahedral arrangements.[4]

Another feature of muscovite that may be observed during heating is also common with the much finer clay mineral particles.[4] Selected specimens of micas and chlorite minerals, when heated slowly above 600°C (1112°F), the temperature coincident with the loss of hydroxyl ions, exhibit exfoliation perpendicular to the plates.[5] Heating vermiculite mica causes sufficient exfoliation to allow the product to be used as a fill insulation. If the micaceous specimens are heated to a sufficiently high temperature, the stacks will collapse and fuse. However, specimens removed from the furnace before complete dehydroxylation and exposed to the humidity of room atmo-

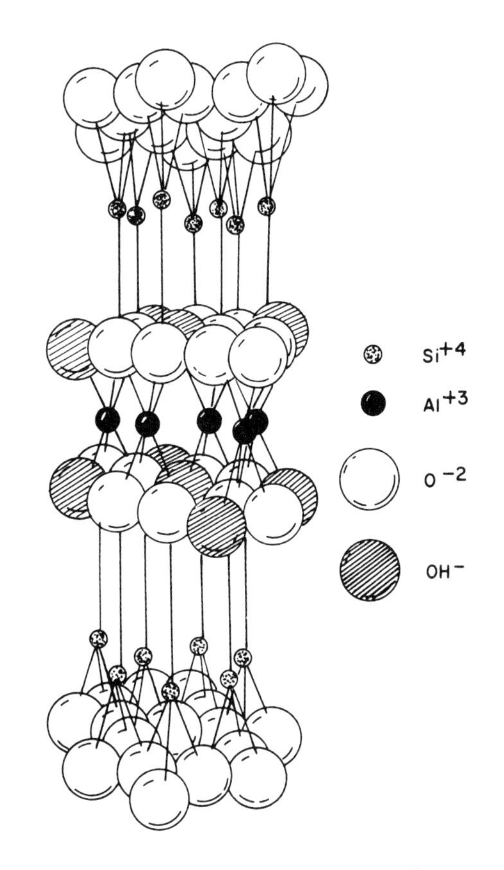

Fig. 1–3 Structure of pyrophyllite, $Al_2(Si_2O_5)_2(OH)_2$. Montmorillonite has the same structure, except some substitution may occur, such as Mg^{+2} or Fe^{+2} for Al^{+3}.[4]

sphere collapse instantly.[6] Specimens stored in a dry atmosphere maintain their exfoliated condition.

Even though it is impossible to observe this type of behavior in clays with the naked eye, there is substantial evidence to indicate that most clay minerals act in a similar manner during heating. The clay particles expand slightly during the early stages of heating and then collapse, causing a shrinkage immediately following dehydroxylation.[7,8] This is the source of a number of problems in fabricating ware which will be discussed throughout the book, but the solution to these problems is simplified by visualizing mica as a model for the platelike clay minerals.

Glasses

Glass has been defined as an inorganic substance in a condition which is continuous with and analogous to the liquid state of that substance but which, as a result of a reversible change in viscosity during cooling, has attained so high a degree of viscosity as to be for all practical purposes rigid.[9] In simple terms glass is really a liquid which when cooled becomes so rigid that it acts as a solid.

The high viscosity attained by a glass melt as it is cooled prevents the atoms from moving about with sufficient freedom to arrange themselves in their most comfortable position, that of a repeating crystal structure. A random, nonrepeating structure is frozen into the glass. This is sometimes called amorphous, glassy, liquid or noncrystalline.

To produce the random glass structure, certain criteria are necessary as far as the atomic arrangement is concerned; these have been explained by Zachariasen.[10] One such condition is that the anions (negatively charged ions) must not be linked to more than two cations (positively charged ions). If more than two cations are bonded to one anion, the rigidity of the bonding prevents the distortion necessary for the random glass arrangement.

In crystalline quartz, SiO_2, the Si-O-Si bond angle is 180° (Figure 1–4). In other words, the two Si^{+4} ions associated with each O^{-} ion are diametrically opposite each other.[11] They try to get as far apart as possible because of their high positive charge and the fact that similar charges repel each other.

In molten silica glass, because of the high temperature, the ions are in a state of constant motion. During the cooling process, the ions try to arrange themselves into the orderly, repetitive crystal arrangement of Figure 1–4 A, but because the viscosity of the liquid silica melt increases so rapidly when cooled, it becomes impossible for the ions to move with sufficient freedom. Thus, a random arrangement is frozen in. The resulting arrangement is that of Figure 1–4 C. The Si^{+4} ions are not diametrically opposite each other, and

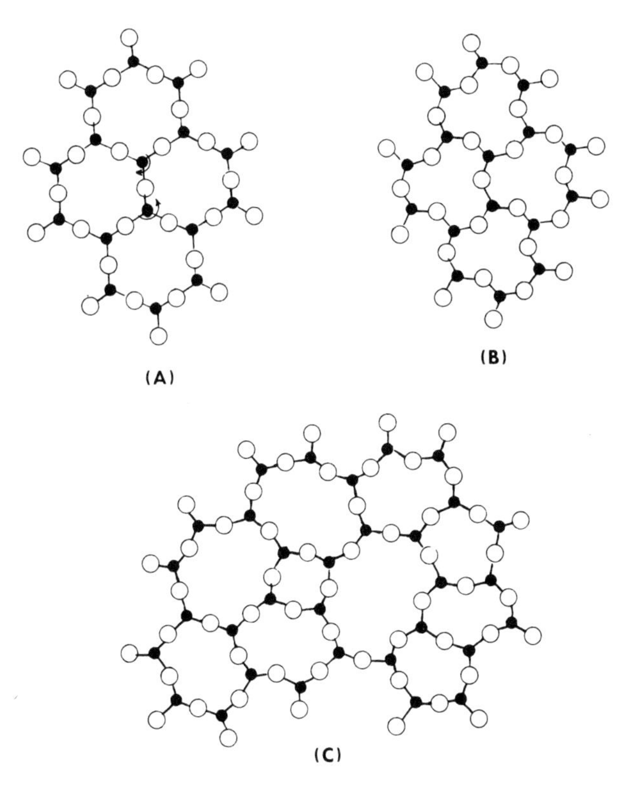

Fig. 1–4 Two-dimensional representation of (A) crystalline, (B) silica and (C) silica glass. In the case of quartz, the low-temperature form (A) may change to the high-temperature form (B) by the simple rotation of the tetrahedra as indicated.

this results in a nonrepeating, random structure characteristic of all glasses. Crystalline quartz glass might be described as a three-dimensional random network of SiO_4^{-4} tetrahedra. It should also be noted that because of the random nature of the glass structure, there are no two points that are identical or equivalent. There are varying degrees of weakness where flow can initiate. Consequently, a glass softens gradually rather than having the sharp melting point characteristic of a crystalline material.

Because of the high softening point and viscosity of pure silica glass, it becomes necessary to make additions to lower its melting

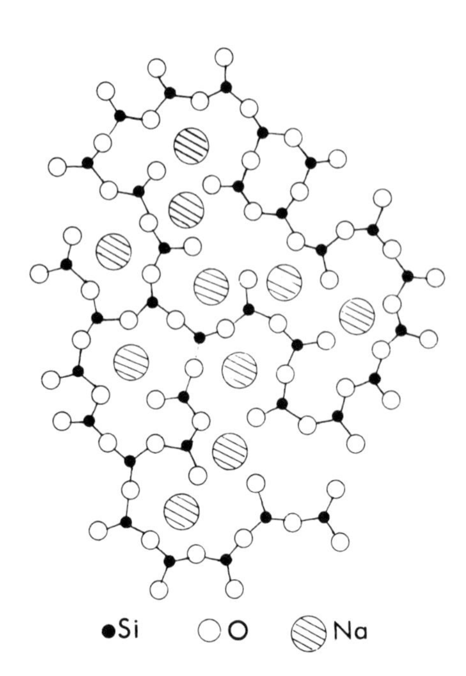

Fig. 1–5 The soda-silica glass structure. Each silicon is tetrahedrally coordinated with oxygens. The fourth oxygen is not shown because it would be above or below the plane of the paper. Oxygens bonded to only one silicon are shown and form discontinuities in the silicon-oxygen network, which weaken the structure. The large sodium ions fit into holes in the network.

point and working range to a more practical working level. To this end, fluxes such as Na_2O, K_2O, CaO, MgO, BaO, and B_2O_3 are added. Simply stated, the addition of these oxides increases the oxygen:silicon ratio above the 2:1 ratio present in pure silica glass. As a result, the silicon-oxygen network becomes discontinuous, breaks or weak points in the structure result, melting points are reduced, viscosity is decreased and thermal expansion is increased. This is illustrated in Figure 1–5.[12]

REFERENCES

1. Hauth, W. E., "Crystal Chemistry in Ceramics," *Bull. Am. Ceram. Soc.*, January–June, 1951.
2. West, R., Gould, R. E., Lux, J. F. and Coffin, L. B., "A New Milling Process: I. Produce Better Whiteware with High Intensity Dispersion;

II. Why High Intensity Dispersion Does the Job So Well," *Ceram. Ind.*, 71, 3: 124–31 (1958).
3. Kaolin, Bulletin No. TD67, Georgia Kaolin Company, 1978.
4. Weaver, D. S. "Identification of Micaceous Minerals in Clay," B.S. Thesis, Alfred University, 1957.
5. West, R., "Solving Firing Problems with Differential Thermal Analysis," *Ceram. Age,* 72, 3: 14–16, 42 (1958).
6. Roy, R. "Decomposition and Resynthesis of the Micas," *J. Am. Ceram. Soc.*, 32: 202–209 (1949).
7. West, R. "Dilatometry of Structural Clay Products," *J. Can. Ceram. Soc.*, 1965.
8. ———, Fleischer, D., Hecht, N., Hoskyns, W. R., Muccigrosso, A. and Schelker, D. H., "Causes of Chipping of a Stiff-Mud Facing Brick," *J. Am. Ceram. Soc.*, 43, 12: 648–54 (1960).
9. Morey, G. W., *The Properties of Glass,* 2nd ed., Reinhold Publishing Corp., N.Y., 1954.
10. Zachariasen, W., *J. Am. Chem. Soc.*, 54: 3941 (1932).
11. Warren, B. E., "Structure of SiO_2 Glass," *Z. Krist.*, 86: 349 (1933).
12. ——— and Biscoe, J., "Fourier Analysis of X-Ray Patterns of Soda-Silica Glass," *J. Am. Ceram. Soc.*, 21: 287–93 (1938).

CHAPTER 2

THE NATURE OF CLAYS

Clays form a unique group of materials. They differ from other materials because of their behavior when associated with water. They develop plasticity and can be shaped by a variety of methods. When dried, shrinkage occurs and strength develops. It is small wonder that such an unusual material has provided the essential ingredient in ceramic bodies for thousands of years.

Clay may be defined in several ways. The chemist or mineralogist might define it as a hydrated alumina-silicate. Ceramists have adopted the following definition: "Clay is a fine-grained rock which, when suitably crushed and pulverized, becomes plastic when wet, leather hard when dried and on firing is converted to a permanent rocklike mass."

There are a great number of clay minerals, and many types of classification may be used. For our purposes we shall limit our discussions to two clay minerals, kaolinite and montmorillonite, inasmuch as they are the significant constituents of the clays used by potters. Because clays were formed by a geological weathering process acting on granites and other rock masses, they do not have definite compositions. Clays usually contain one or more clay minerals, as well as accessory minerals such as quartz, muscovite, biotite, ilmenite, leucoxene, anatase and hydrous iron oxides.

Current mining and beneficiation techniques are capable of separating a large portion of these accessory minerals from the mined clay. Sedimentation processes are used which easily separate particles of different size and specific gravity. The clay suppliers can provide products of high purity, specific particle size ranges and various degrees of crystallinity.

Pure kaolin has a refractoriness approaching 1780°C (3236°F), but, if impurities are present, the softening temperature will be reduced and other properties altered. Following are some of the effects produced by accessory minerals.

Silica

Silica occurs in clays in the free state as crystalline quartz or in the amorphous form as hydrated or colloidal silica. It is present in most clays because it is a constituent of igneous rocks and is unaltered during the weathering process.

The effects of free silica are a reduction in plasticity, drying and firing shrinkage, fired strength, refractoriness and an increase in thermal expansion.

Alumina

Combined alumina occurs in feldspars, mica, hornblende, tourmaline, bauxite, laterite, gibbsite, diaspore and the clay minerals. Accessory minerals containing alumina will reduce plasticity and increase refractoriness.

Alkalies

Alkali compounds are usually associated with alumina compounds and have great effects on properties. Feldspars, micas or hydrous micas are the principal alkali-bearing accessory minerals. Alkalies may also be present as a result of adsorption of these ions on the surface of clay minerals. Alkalies also occur in clays as soluble salts, such as sulfates and chlorides.

Alkalies, because of their strong fluxing action, reduce the refractoriness and vitrification temperature of the clay. Because of the increase in glass formation, the strength of the fired body is increased and absorption decreased. The presence of alkalies may change the plastic properties of the clay, particularly if they are soluble alkalies. A clay containing alkali will not be as plastic as one without it. It will not retain its shape after forming.

Iron Compounds

Several varieties of iron oxides, sulfides, carbonates, hydroxides and silicates may be found in clays. Magnetite (Fe_3O_4), hematite (Fe_2O_3), limonite ($Fe_2O_3 \cdot xH_2O$), goethite ($Fe(OH)_3$), ferrous oxide (FeO), pyrite (FeS_2), ferrous sulfate ($FeSO_4 \cdot 7H_2O$) and siderite ($FeCO_3$) are common examples.

The iron oxides are very susceptible to oxidation-reduction reactions and because Fe_2O_3 and FeO are very different in their effects, awareness of these differences is necessary. The Fe_2O_3 does not greatly reduce the refractoriness of a clay and Fe_3O_4 in combination with silica starts glass formation at a temperature of 1455°C (2651°F). But if reduction occurs and produces more FeO, glass formation can start at approximately 1180°C (2156°F). The Fe_2O_3 is red in color and is responsible for the various shades of the brick red colors. Ferrous oxide is black and will react with silicates and alumina-silicates to form compounds melting in the 1100°C (2012°F) range. Proper control of the furnace atmosphere is an important procedure in the firing of clays containing iron oxide.

Clays containing FeS_2 also require special attention during the firing process. Pyrite is oxidized to Fe_2O_3 in the 400 to 600°C (752 to 1112°F) range with the evolution of SO_2 gas. If oxidizing atmospheres are maintained and sufficient time given for the decomposition to take place, the presence of sulfide compounds is of little harm to the final product. If, however, reducing conditions are present which prevent the decomposition of sulfides and also result in the formation of FeO, serious problems may arise. Because of the powerful fluxing action of FeO, glass formation may result before decomposition of the sulfide is complete. This can readily result in bloating.

Calcium Compounds

Calcium minerals such as calcite, aragonite, $CaCO_3$, and gypsum, $CaSO_4 \cdot 2H_2O$, are found in many clays. They decompose to form calcium oxide, CaO, which acts as a flux. The resulting glass formation reduces the vitrification temperature, resulting in an increase in fired strength and decreased absorption. By combining with the iron minerals, CaO bleaches the red color to buff. If not fired to a sufficiently high temperature, the CaO may remain as free lime. After firing, this will react with water vapor to form calcium hydroxide, $Ca(OH)_2$, and large expansion and sufficient pressure can be developed to disrupt the fired product. This is lime popping. It is necessary to fire clays containing calcium compounds sufficiently high to cause the CaO to react with the other ingredients of the body. Such reaction results in the formation of minerals such as anorthite, $CaO \cdot Al_2O_3 \cdot 2SiO_2$, as well as glass. The CaO is rendered completely insoluble and unreactive.

Carbonaceous Materials

Carbonaceous materials, such as peat, lignite and coal, are present in some clays, which are usually dark in color and have an off-white fired color. Large amounts of these clays will promote reducing conditions in the kiln. Color, vitrification, decomposition and oxidation-reduction reactions are affected by the presence of inordinate amounts of carbonaceous materials.

The amount and type of accessory material or minerals present in a clay often determine its usefulness. Several clays have names based upon their geological history and the accessory minerals present.

Residual kaolins are found near the original parent rock. The main deposits in the United States are in North Carolina. In England, English China clay is a typical example. The methods of mining and beneficiation remove the associated unweathered rock and sand. These clays are characteristically coarse grained and exhibit low plasticity.

Sedimentary kaolins, as the name implies, have been transported by water and redeposited at some lower level. These deposits have been classified by nature in the transport process. They are fine grained and have good plastic properties. The clays of South Carolina, Georgia and Florida are of this type.

Ball clays are a special group of sedimentary kaolins. They were redeposited in low-lying swampy areas and therefore are usually associated with considerable amounts of organic matter. They are finer grained than the other sedimentary kaolins and are characterized by a high degree of plasticity and dry strength and high drying shrinkage.

Fire clays are classified into three types.[1] *Flint* fire clays are hard, exhibit conchoidal fracture and show little plasticity, regardless of how fine they are ground. When mixed with a plastic clay they serve as grog which reduces shrinkage. Flint clays are abundant in the United States. Their composition approaches kaolin and they fire to a white or light cream color. *Plastic fire* clays show great variation in composition. They are categorized as low or high heat duty, depending on their composition or fusion point. The low heat duty clays contain comparatively more flux, and their use is limited to ladle brick, mortars and castables. They vary greatly in silica content, and those high in silica are used to produce a desirable high silica fireclay brick. *High alumina* fire clays are used in the manufacture of super duty high alumina refractory brick.

Brick clays encompass a wide variety of clays associated with considerable amounts of fluxes and coloring oxides. These are useful in producing products that mature at a moderate firing temperature.

The main mineral constituents of brick clays and shales are chlorite, illite, kaolinite with accessory minerals of quartz, mica, and various iron-bearing minerals. The fired color may be variable. The high lime clays and shales are buff burning while the common brick salmon to red color is due to the FeO present.

Stoneware clays are composed of fine-grained kaolinite and accessory minerals, such as feldspar, rutile, silica and iron-bearing minerals, which provide the fluxing necessary for vitrification to a dense impervious body. Prepared stoneware bodies have replaced natural clays. Uniformity and duplication of results are more dependable.

Bentonite is produced by the weathering of a volcanic glass. The mineral constituent is montmorillonite. Two varieties are common in the United States, the Western or swelling variety, and the Southern or nonswelling variety. Both are characterized by their extremely small particle size, which results in large drying shrinkage and high dry strength. Their main use is as small additions to improve plasticity, strength or as a suspending agent in glazes or enamels.

The Size of Clay Minerals

Figures 2–1 and 2–2 are electron microscope pictures of kaolinite particles. Two main points are evident. They are small and they take the form of hexagonal plates. The ratio of the long dimension to the plate thickness is approximately 10:1. This factor becomes important in later discussions involving the orientation of particles during the forming processes.

Kaolins range in size from 100 to 0.1 microns (μ) average particle diameter as determined by sedimentation methods.[2] (1 μ equals 10^{-6} m. or 10^{-4} cm. or 0.0001 cm.)

Owing to the extremely small particle size of clay minerals, surface effects become extremely important and their behavior is governed largely by surface properties. An idea of the amount of surface present in finely divided materials is shown in Table 2–1, which

Fig. 2–1 Kaolinite crystals. 19,500×. (Courtesy C. E. Randall)

shows the amount of surface generated as one starts with a 1 cm. cube and subdivides it into smaller and smaller cubes.

Thus, 2.61 g. of kaolinite having a size of 0.01 μ would have over 6000 ft.^{2} of surface area. A 100 lb. bag of bentonite would contain approximately 2400 acres of surface. Surface properties are most important in dealing with and understanding the nature of clays. Most of the observed properties of clay-water systems are a function of surface area.

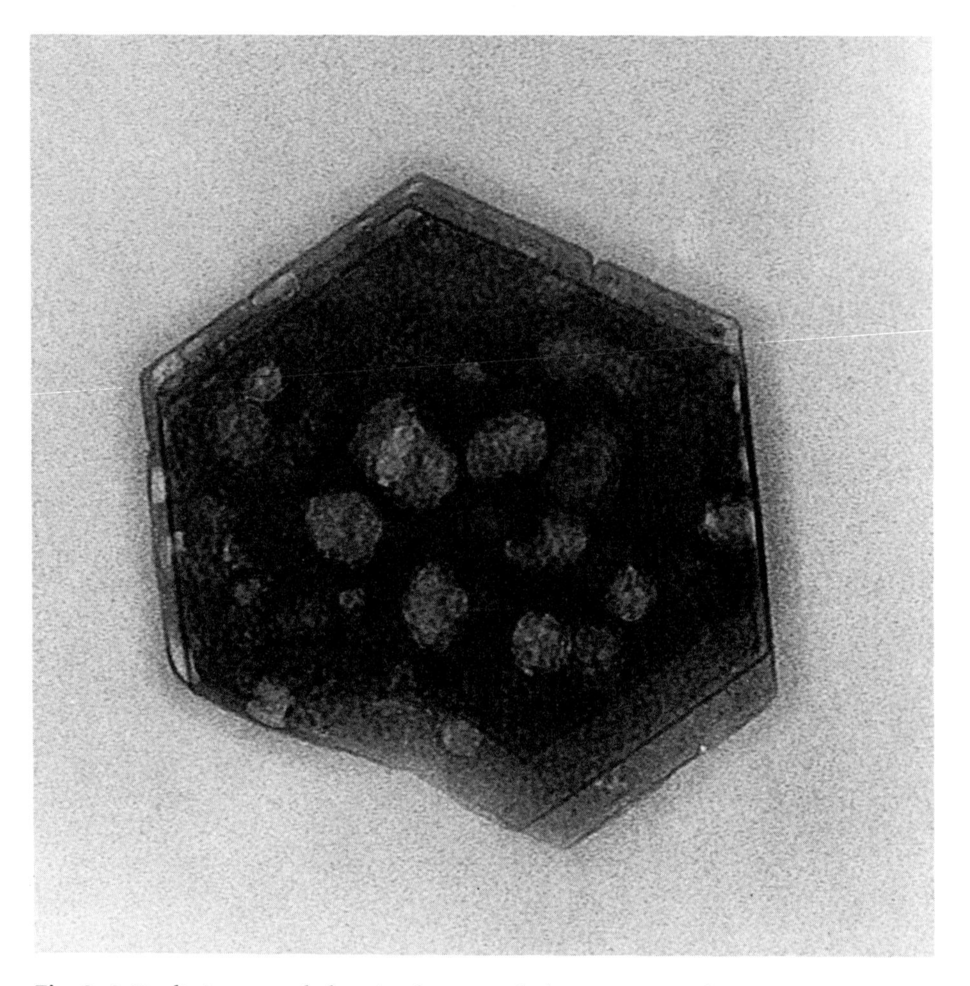

Fig. 2–2 Kaolinite crystal showing hexagonal plate structure. (Courtesy C. E. Randall)

Figure 2–3 shows the particle size distribution curves for four commercial kaolinites. These show a wide range in particle size. Inasmuch as the particle size of clays is in the sub-sieve range (a 325 mesh sieve has an opening of 44μ), the measurement of size distribution is made by the use of some type of sedimentation in air or

TABLE 2–1
SURFACE AREA PRODUCED BY SUBDIVISION OF A CUBE[2]

Length of Edge		Typical Object	Number of Cubes	Total Surface
1 cm.	= 0.3937 in.	Small peas	1	0.93 in.^{2}
1 min.	= 0.0394 in.	Fine shot	10^3	9.30 in.^{2}
0.1 mm.	= 0.0039 in.	Talcum powder	10^6	93.00 in.^{2}
0.01 mm.	= 0.0004 in.	Amoeba	10^9	6.46 ft.^{2}
1μ	= 0.001 mm.	Small bacteria	10^{12}	64.58 ft.^{2}
0.1μ	= 0.0001 mm.	Large colloidal particle	10^{15}	645.83 ft.^{2}
0.01μ	= 0.00001 mm.	Large molecule	10^{18}	6458.30 ft.^{2}

liquid, usually water. Table 2–2 gives the particle size analysis of several clays.

Testing Clays

Because of the small size of clay particles, variable composition and similarity in appearance, it is not possible by visual inspection to deduce immediately the suitability of a particular clay for pottery making.

Several simple tests can be run that will answer many important questions. Practically any clay can be used to make some kind of pottery, but the potter must know the clay he works with. This knowledge is often gained as the result of day-by-day experience. The occasion may arise when quick evaluation of a clay for pottery use

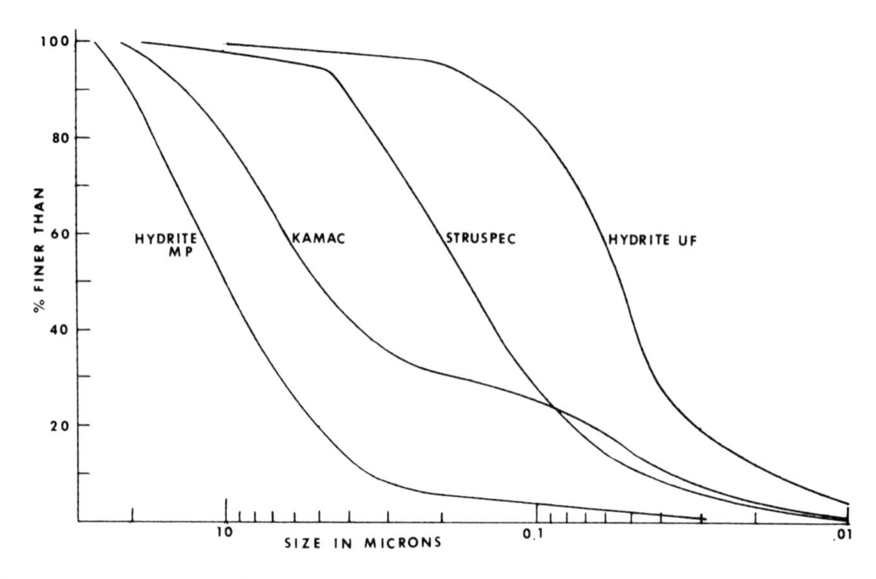

Fig. 2–3 Particle size distribution curves for several kaolinites.

may be necessary. For this purpose certain evaluation tests can be made. Binns[4] states that three properties are demanded of a clay for successful pottery use: plasticity, porosity and suitable vitrification range.

Plasticity is the property that allows one to form a shape. It permits the material to be deformed by pressure and to retain that shape when the pressure is removed. Coarse-grained clays are not as plastic as fine-grained clays. Coarse-grained clays are often said to be short, meaning they cannot stand much deformation without cracking.

A clay may have a high degree of plasticity but be impossible to dry without warping and cracking, because the water cannot easily escape during drying. Porosity of the clay controls the migration of water during drying. Porosity can be improved by sand or grog additions. Fine-grained clays have low porosity and high plasticity, while coarse-grained clays have high porosity and low plasticity. There is always a compromise between these two properties.

TABLE 2–2

PARTICLE SIZE ANALYSIS OF CLAYS[3]

	PERCENT FINER THAN IN MICRONS								
Clay	**20μ**	**10μ**	**5μ**	**3μ**	**2μ**	**1μ**	**0.5μ**	**0.3μ**	**0.2μ**
M and D Ball Clay	99.5	99.1	97.1	94.8	94.1	84.5	67.4	50.0	42.4
No. 17 Fayle Blue Ball	100.0	99.4	98.7	96.0	90.8	76.3	49.5	26.5	10.5
Hydrite UF	100.0	100.0	99.1	98.2	96.5	82.6	42.5	16.5	6.0
Old Hickory No. 5 Ball	98.6	97.3	92.5	86.3	79.3	65.0	37.5	16.5	5.5
Barden Clay	99.5	99.2	97.6	96.8	92.0	79.2	42.5	15.0	4.0
Struspec	100.0	99.3	95.8	77.5	57.5	28.0	11.2	6.0	3.5
NLB English China	100.0	100.0	90.3	74.2	62.9	43.6	22.5	8.0	3.3
Old Mine No. 4 Ball	98.7	96.2	91.0	82.1	74.4	56.5	33.4	15.5	3.0
Lampin Clay	97.5	90.9	77.5	63.5	55.4	37.0	21.5	9.0	2.5
Ajax P	100.0	99.3	98.0	94.7	90.8	68.4	37.0	13.0	2.5
Champion Challenger	99.2	96.8	94.4	90.3	87.1	72.6	43.5	8.0	2.5
Kaolex SC	98.2	91.8	80.0	67.3	56.4	36.5	16.5	5.5	2.0
Bandy Black	98.3	93.2	82.0	71.0	60.0	32.0	21.0	8.5	1.5
Kamac	99.3	79.0	50.0	37.0	31.1	26.0	15.2	7.7	1.4
Pioneer Airfloat	100.0	100.0	94.0	79.4	66.3	42.5	3.3	2.2	1.2
Hydrite MP	90.6	52.0	19.5	10.0	7.0	3.9	1.7	0.7	0.0
Kaolex SH	100.0	99.1	92.4	79.2	64.2	41.5	16.0	5.6	0.0

When a clay is fired, it becomes hard and strong due to glass formation. Firing results in densification and particle-to-particle adherence. It is essential to know the temperature range over which vitrification occurs and the amount of shrinkage that takes place during the firing process.

Knowing all there is to know about a material makes it a servant in one's hands, but lack of this knowledge causes frequent disappointments. The following simple tests will make the potter better acquainted with his clay.

Water of Plasticity

This test determines the amount of water required to develop the optimum plastic properties of a clay. The conduct of this test is illustrated in Figure 2–4. Weigh out 100 g. of dry, pulverized clay. Place it on a glass slab or other nonabsorbent surface. Fill a 100 cc.

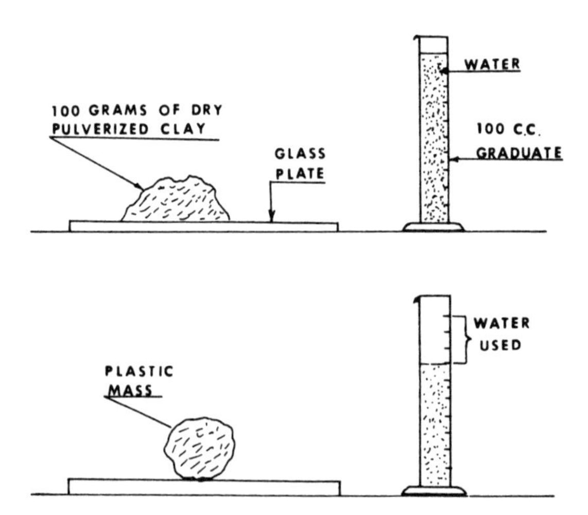

Fig. 2–4 Test for water of plasticity. By starting with 100 grams of clay, the percent water of plasticity is equal to the cubic centimeters of water used to develop maximum plasticity.

graduated cylinder with water. Slowly add water to the clay and mix thoroughly between each addition. Note the water required to develop initial plastic properties, as well as the amount that develops stickiness. This provides the range of water content in which suitable workability is evident. The cubic centimeters of water used equals the percent water of plasticity (1 cc of water weighs 1 g.). A fine-grained plastic clay may require 40 percent water and a coarse-grained nonplastic clay only 25 percent.

Drying and Firing Shrinkage

From the plastic mix make a shape approximately 12 cm. long by 2 cm. wide by 1 cm. thick, as illustrated in Figure 2–5. This can be done by hand molding or rolling and cutting out the shape desired. Make two marks on the top surface of the bar exactly 10 cm., or 100 mm., apart. Dry the bar and measure the distance between the marks. If the original distance between the marks was 100 mm. and the distance after drying was 95 mm. the drying shrinkage was 5 percent.

By firing to the vitrification temperature and remeasuring the distance between the marks, the firing shrinkage can be determined. If the distance between the marks is now 85 mm., the total drying plus firing shrinkage is 15 percent, of which 5 percent is drying shrinkage and 10 percent firing shrinkage. It is necessary to fire at several temperatures in order to obtain a complete picture of shrinkage behavior.

Absorption

The fired piece is weighed to the nearest 0.1 g. It is then placed in a suitable container and boiled for five hours in water. It is removed from the water, the excess wiped off, and the piece reweighed. The percent absorption is equal to:

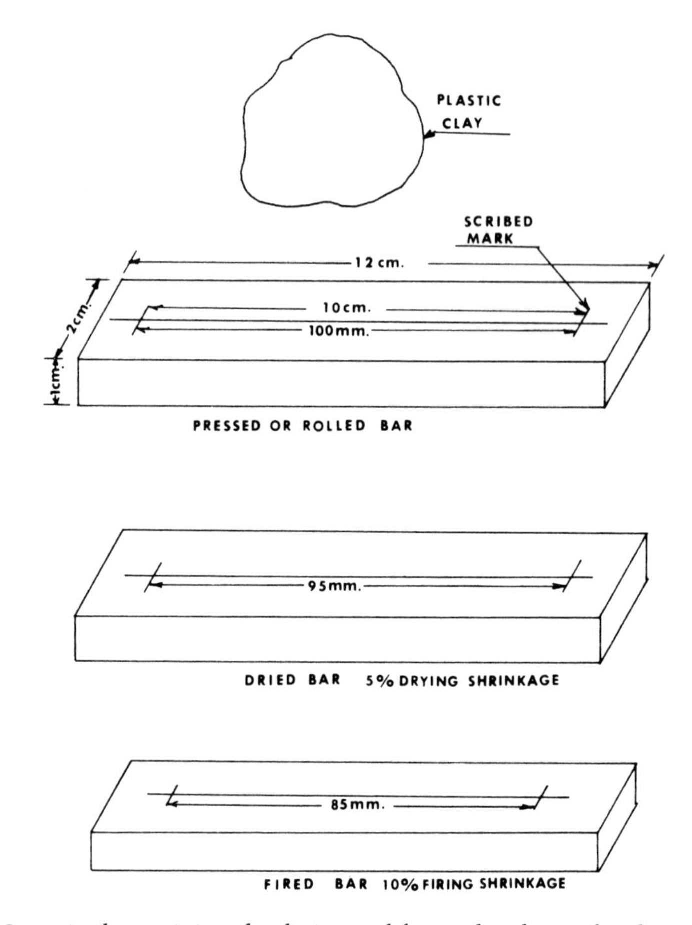

Fig. 2–5 Steps in determining the drying and firing shrinkage of a clay.

$$\frac{\text{Final Weight} - \text{Original Weight} \times 100}{\text{Original Weight}}$$

as illustrated in Figure 2–6.

To obtain a complete picture of the effect of different temperatures on shrinkage and absorption, several firings must be made. The effect of firing temperature on the absorption and firing shrinkage of plastic, medium plastic and short red-burning surface clays is shown in Figures 2–7 and 2–8.

$$\% \text{ ABSORPTION} = \frac{\text{GAIN IN WEIGHT}}{\text{ORIGINAL WEIGHT}} \times 100$$
$$= \frac{5}{50} = 10\%$$

Fig. 2–6 Steps in determining the percent absorption of a clay specimen.

The combination of absorption and firing shrinkage data provides the basis for determining the optimum firing range of a clay, as shown in Figure 2–9.

If one is going to do a considerable amount of clay testing, it may be worthwhile to consider making a thermal gradient furnace. This is a firing chamber in which planned temperature differences exist from one part to another. Such a furnace can be made in several ways. A simple one utilizes the fact that a resistance heating element, such as Globar, is naturally hottest in the middle and coolest at the end. If a furnace similar to the one in Figure 2–10 is made, a natural temperature gradient of approximately 300°C (572°F) will exist. Associated with any thermal gradient furnace is the necessary mechanism for determining what the temperature is at various spots along the length of the specimen being fired. In the Globar furnace, a rack and pinion device moves the thermocouple along the length of the specimen. In this way a temperature profile can be determined.

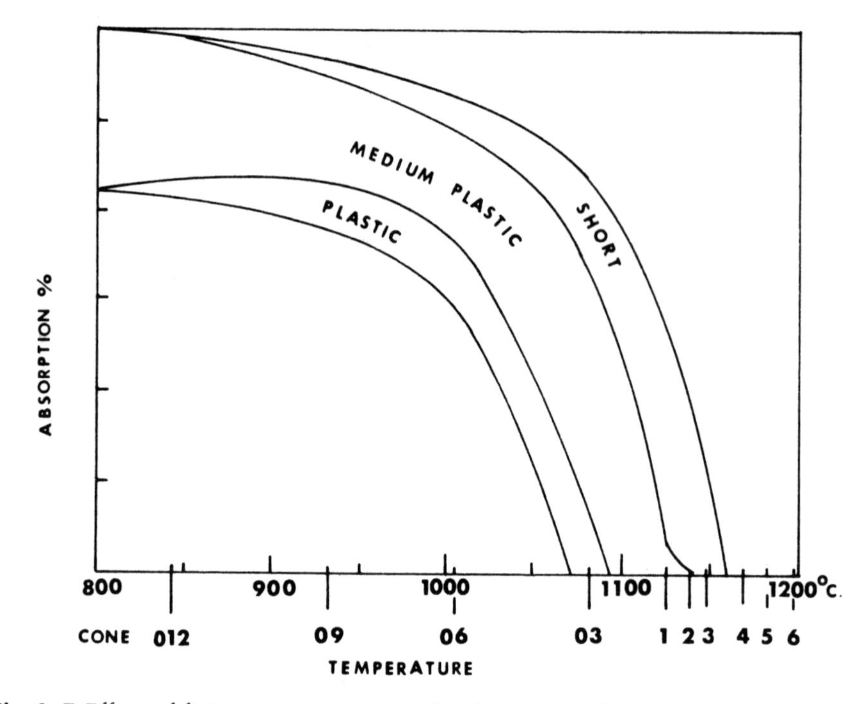

Fig. 2–7 Effect of firing temperature on the absorption of plastic, medium plastic and short red firing surface clays.

A more simple thermal gradient furnace can be made from any available furnace, either electric or gas, by building a simple temporary muffle inside the firing chamber. By having one end of the muffle near the center of the firing chamber and the other end bricked into the door of the kiln, a temperature gradient will be produced. Such an arrangement may be laid up on a setter slab using fire clay splits as shown in Figure 2–11. Its location and the layup of the brick in the door should be consistent. A mechanism for determining temperatures along the length of the specimen is essential and the position of the thermocouple bead with respect to the specimen must be known and reproducible. Methods of making the temperature survey can be worked out to the satisfaction of the individual. Mechanical movement of the thermocouple is more dependable, more ac-

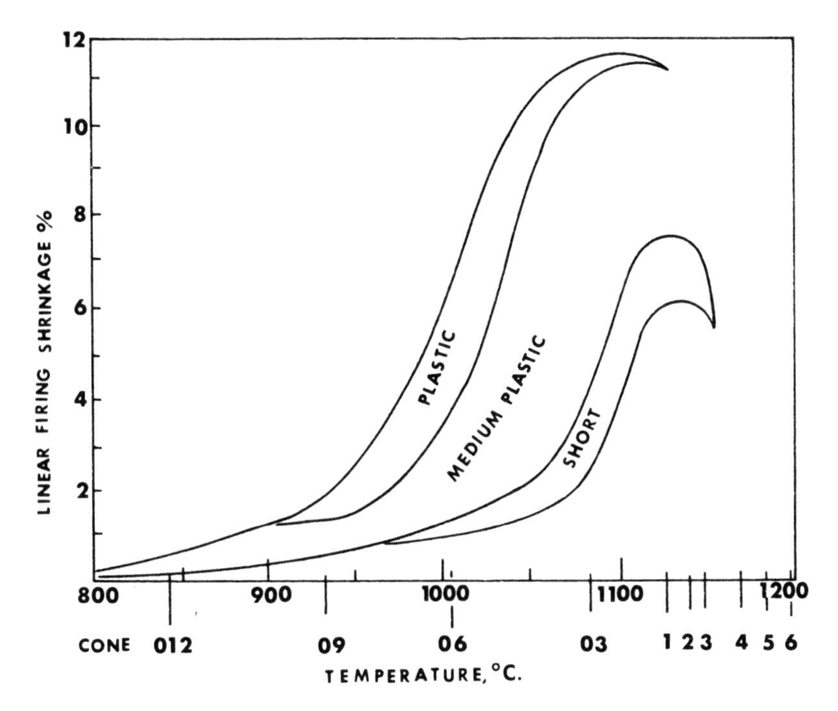

Fig. 2–8 Effect of firing temperature on the linear firing shrinkage of plastic, medium plastic and short red firing surface clays.

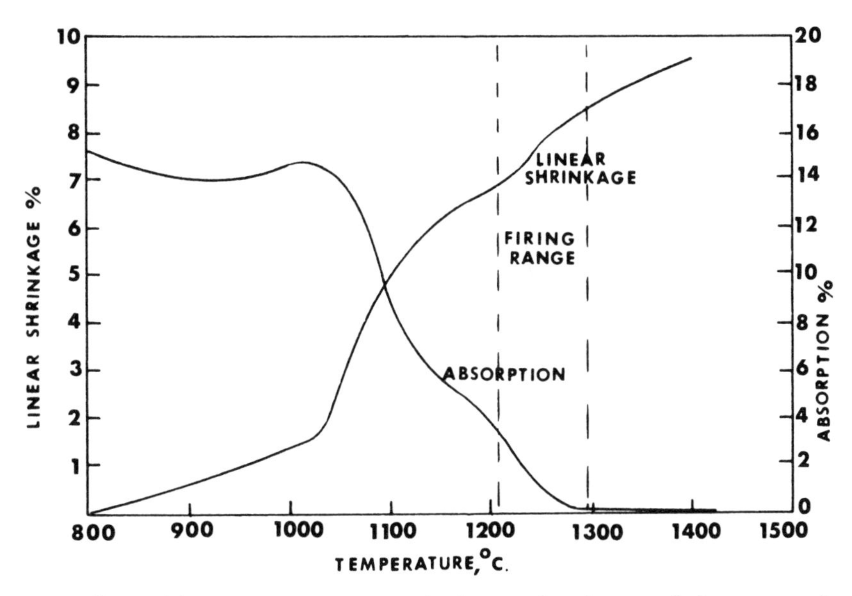

Fig. 2–9 Effect of firing temperature on the linear shrinkage and absorption of a typical stoneware clay. The probable suitable firing range is indicated.

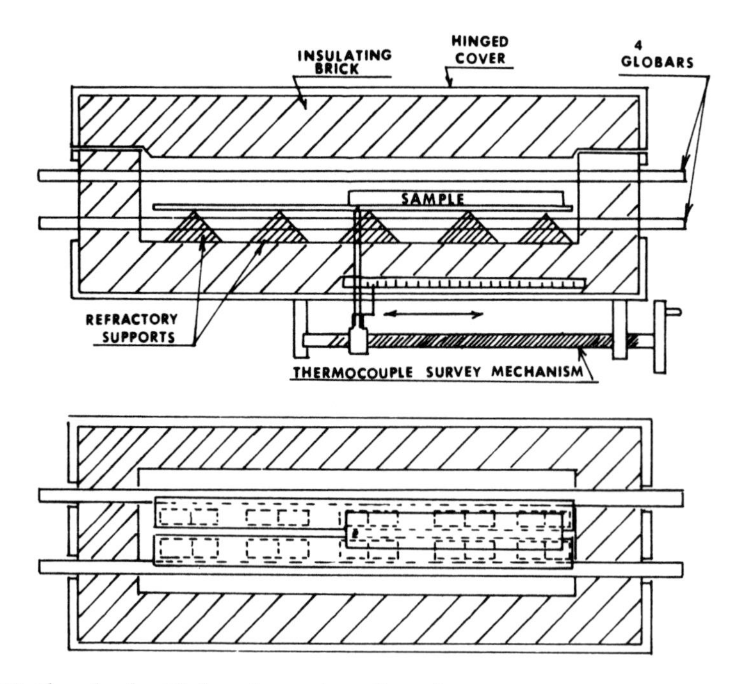

Fig. 2–10 Sketch of a Globar thermal gradient furnace.

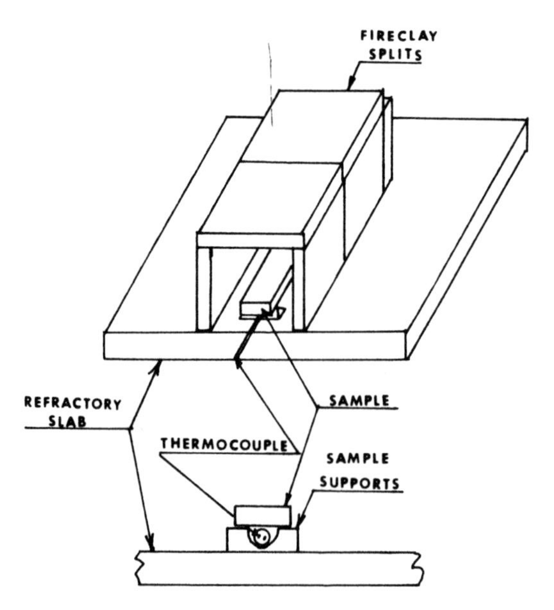

Fig. 2–11 Sketch of a low-cost thermal gradient furnace made by construction of a muffle that may be inserted in the firing chamber of a regular kiln.

curate and easier on the thermocouple protection tube than is hand manipulation. If money is no object, ready-made thermal gradient furnaces are available.

The specimen from a thermal gradient furnace is illustrated in Figure 2–12. If the temperature distribution along the long axis of the specimen is known, the effect of temperature on many properties can be determined in a single firing. Shrinkage can be measured directly. The proper firing range as well as the fired color can be observed. By slicing up the specimen, densities and absorptions can be measured as a function of temperature.

The use of this technique in studying glazes has obvious advantages.

Differential Thermal Analysis[5]

It is well known that every chemical reaction either gives off or takes on heat. The evolution of heat is an exothermic reaction. Typical of this type of reaction is the oxidation of carbon: $C + O_2 \rightarrow CO_2 +$ heat. The type of reaction that requires heat to make it proceed is an endothermic reaction. In general, reactions resulting in decomposition or breakdown of a material are endothermic; for example, calcium carbonate, $CaCO_3$, $+$ heat $\rightarrow CaO + CO_2$. The loss or evolution of heat will raise the temperature of the sample with respect to its surroundings. If heat is required for the reaction, the temperature of the sample will lag behind that of its surroundings. These facts allow one to determine the temperatures at which a reaction or change takes place in a sample during heating. By observing the magnitude of the gains or losses of heat and the temperature at which these occur, one can distinguish one clay mineral from another.

The essential parts of the differential thermal analysis apparatus are a differential thermocouple, sample holder, furnace, recorder and a means of raising the temperature of the furnace at a reasonably uniform rate.

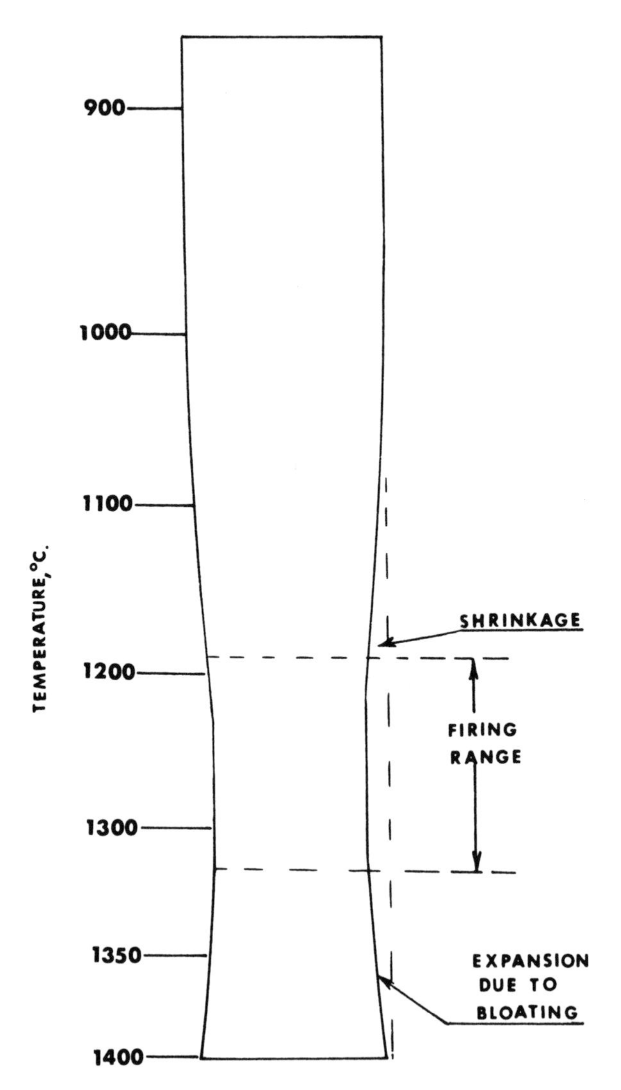

Fig. 2–12 Sketch of a fired thermal gradient specimen showing shrinkage, firing range and expansion due to bloating.

A thermocouple consists of two dissimilar metal wires which, when welded together, generate a small but measurable electric potential that is a function of the temperature at the junction of the two wires. Several types of metal wires may be used, depending on the temperature to be measured. The most common pairs are copper-constantan, chromel-alumel and platinum with platinum-10% rhodium. Standard charts are available giving the voltage generated as a function of temperature.

A single thermocouple and differential thermocouple are illustrated in Figure 2–13. The single couple will measure the temperature at T_1. The differential couple will measure the temperature difference between T_1 and T_2. This provides the basis for the differential thermal analysis measurement.

The next requirement is a holder that contains the sample to be tested around one thermocouple junction and an inert sample around the other. An inert sample is one that undergoes no changes when heated. Calcined alumina is usually used for this purpose. A convenient method of providing the container for low-temperature work (below 1050°C or 1922°F) is to obtain a nickel block and drill two holes in it. These serve as the sample holding cavities. Two additional holes are then drilled perpendicular to the sample cavities. These holes intersect the sample cavity holes and allow for the insertion of the differential thermocouple and protection tubing, placed so that

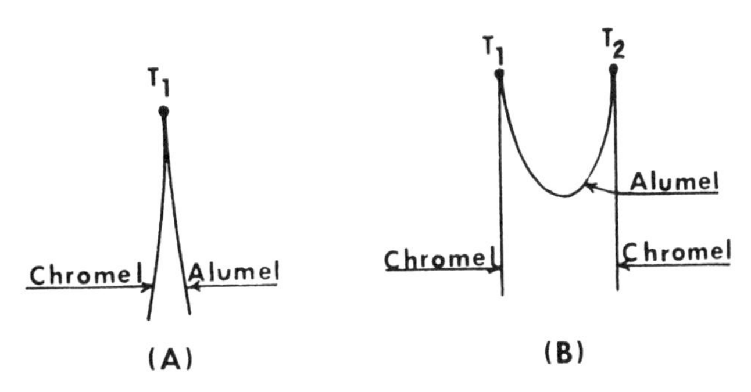

Fig. 2–13 (A) Single thermocouple and (B) differential thermocouple.

the thermocouple beads are in the middle of the sample cavity. Another hole provides for insertion of a single couple that indicates the actual temperature in the center of the assembly (Figure 2–14).

With proper setting of the nickel block on a refractory support which can be inserted into a furnace and a method of recording the block temperature and differential temperature, a thermal analysis apparatus is complete. The apparatus can be rather crude and cheap, or it may be refined by electronic components that will raise the temperature of the furnace at an automatically predetermined rate. X-Y recorders are available which automatically plot the differential thermal analysis curve. Such refinements are necessary when quantitative results are desired. For rough qualitative work involving determination of the type of minerals present or temperatures at which changes are taking place, such refinements are unnecessary.

Clay minerals, when heated, absorb (endothermic reaction) or give off heat (exothermic reaction) at definite temperatures that are characteristic of a particular clay mineral. Therefore, each mineral has its own characteristic thermal analysis curve.[6] Figure 2–15 shows these curves for kaolins, ball clays and bentonite. These curves show many similarities as well as differences.

The differential thermal analysis curve for kaolin is characterized by two peaks, an endothermic peak in the 550 to 650°C (1022 to 1202°F) region and a very sharp exothermic peak at 980°C (1796°F). If these peaks are present, the mineral kaolinite is present in the

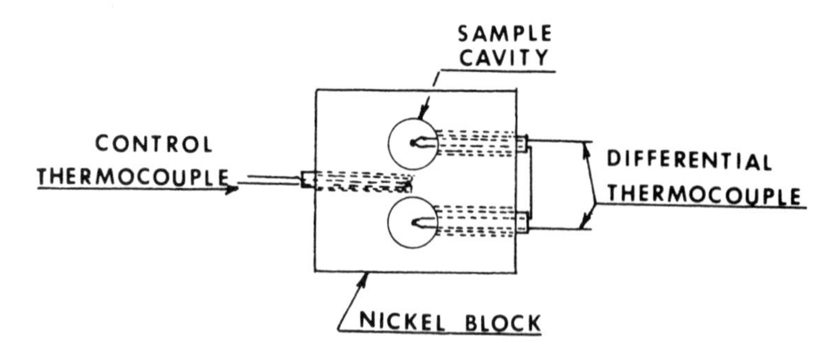

Fig. 2–14 Differential thermal analysis sample block assembly.

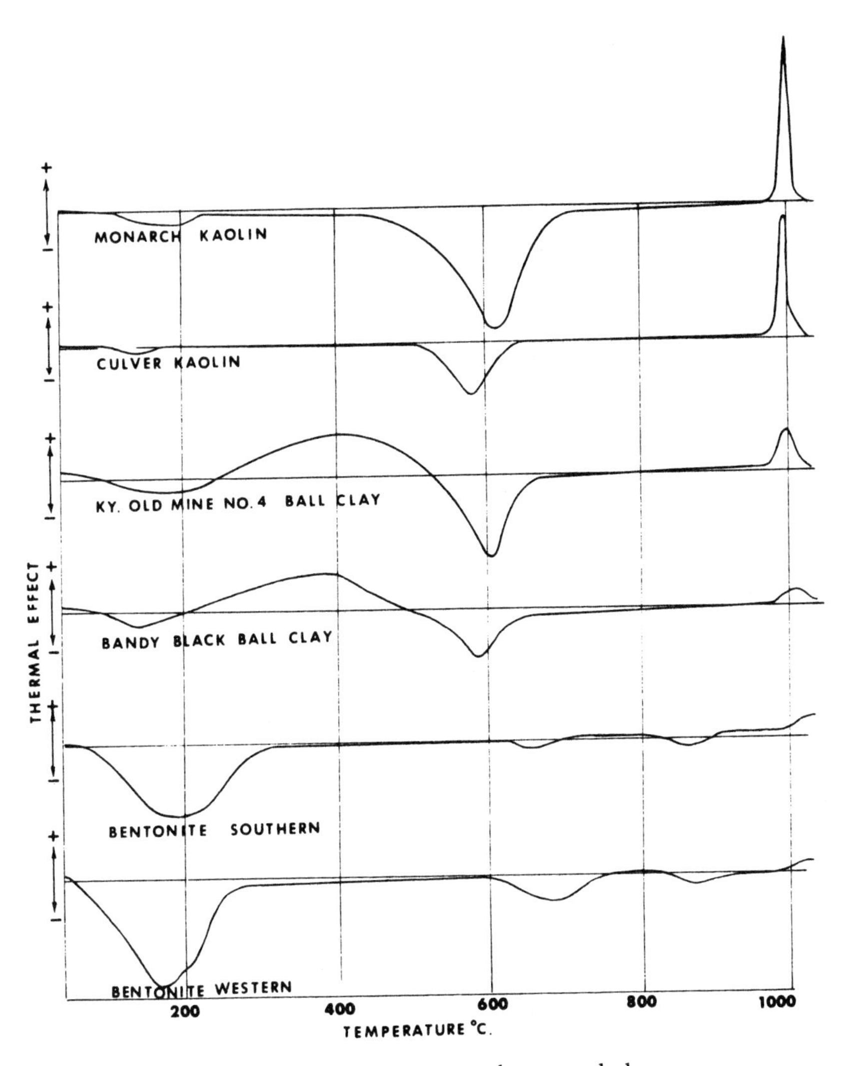

Fig. 2–15 Differential thermal analysis curves for several clays.

sample. We see that the first four curves for the two kaolins and two ball clays have these common peaks and kaolinite is the main constituent. We also observe a small endothermic peak in the 100 to 200°C (212 to 392°F) region. This is due to the heat required to remove water that has been adsorbed on the surfaces of the clay particles. As the surface area increases (smaller particle size), the amount of this adsorbed water will increase. Hence, we observe a very small endothermic effect at this temperature in kaolins, slightly larger in ball clays and a major peak in bentonites.

One other significant difference is noticeable between the kaolins and ball clays: the gradual exothermic effect in the ball clays in the 300 to 500°C (572 to 932°F) temperature range. This is due to the oxidation of organic matter present in most ball clays. This oxidation of carbon evolves heat and the extent of this effect is proportional to the amount of organic matter present in the clay. Such a peak on a thermal analysis curve serves as a warning that, in firing such a clay, this temperature should be traversed slowly with good oxidizing conditions so that the organic matter is completely removed by oxidation to carbon dioxide. If this is not done and the gases are generated later in the firing during glass formation, bloating will result.

The bentonites are characterized by their large endothermic peak in the low temperature region and two small endothermic peaks at approximately 680 and 880°C (1256 and 1616°F). There is complete absence of any exothermic peak in the 1000°C (1832°F) region.

The differential thermal analysis curve tells us at what temperature something is happening, but it does not tell us what is happening. To do this, other techniques, such as X-ray diffraction, must be used. By studying the mineral before and after each temperature where a change occurs, it is possible to learn what has happened. From such data the changes occurring in a kaolin when heated can be deduced.[7,8]

100–200°C (212–392°F) ENDOTHERMIC PEAK

This results from the heat required to drive off the free water absorbed on the surface of the particles. The size of this peak depends roughly on the surface area and types of ions that may be associated with the adsorbed water.

450–600°C (842–1112°F) ENDOTHERMIC PEAK

From the chemical formula of kaolinite, $Al_2Si_2O_5(OH)_4$, and the structural drawings previously shown, it is evident that kaolinite contains $(OH)^-$ ions. As kaolinite is heated in this temperature range the chemical water (hydroxyl) is removed from the kaolinite according to the following reaction:

$$\underset{\text{kaolinite}}{6(Al_2O_3{\cdot}2SiO_2{\cdot}2H_2O)} \rightarrow \underset{\text{metakaolin}}{3(2Al_2O_3{\cdot}4SiO_2)} + \underset{\text{water}}{12H_2O}$$

During this process, there is a weight loss of 13.95 percent. The hexagonal plates of the kaolinite remain the same shape but show no evidence of crystallinity. With this change to metakaolin, the material also loses its plastic properties. It is known that metakaolin can be rehydrated by extended exposure to water to again form kaolinite. In doing so it regains its plastic property. During such rehydration, an expansion takes place corresponding to the shrinkage observed when the water is driven off.

980°C (1796°F) EXOTHERMIC PEAK

At this temperature the metakaolin layers condense to form a new type of crystal structure called a spinel (Figure 2–16). It has the approximate composition $2Al_2O_3{\cdot}3SiO_2$.

$$\underset{\text{Metakaolin}}{3(2Al_2O_3{\cdot}4SiO_2)} \rightarrow \underset{\text{Spinel phase}}{3(2Al_2O_3{\cdot}3SiO_2)} + \underset{\text{Amorphous silica}}{3SiO_2}$$

1050–1100°C (1922–2012°F) EXOTHERMIC PEAK

Slightly above the 980°C (1796°F) exothermic peak, another smaller peak may be observed, which is a result of the transformation of the spinel phase to mullite. This is known as *primary mullite* as compared to *secondary mullite,* which crystallizes from feldspathic glass

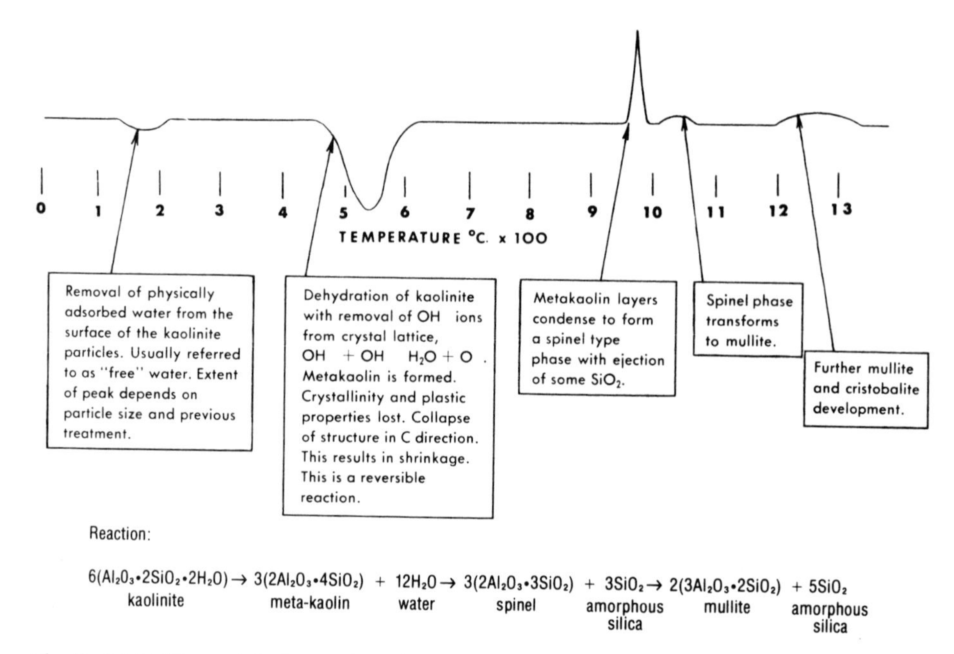

Fig. 2–16 Differential thermal analysis curve for kaolinite indicating the changes associated with each peak.

that may be present in a body. More amorphous silica is ejected during this reaction.

$$\underset{\text{Spinel phase}}{3(2Al_2O_3 \cdot 3SiO_2)} \rightarrow \underset{\text{Mullite}}{2(3Al_2O_3 \cdot 2SiO_2)} + \underset{\text{Amorphous silica}}{5SiO_2}$$

Above 1100°C (2012°F), there is continued development of mullite and cristobalite. If low-cristobalite bodies are desired, the length of time the body is exposed to this temperature range should be as short as possible.

Fired Color of Clays[9–14]

The fired whiteness of a ceramic body has little significance for some types of ware, such as sanitary ware or electrical porcelain, mainly because opaque glazes are employed. However, the whiteness of dinnerware and porcelain artware is probably one of their most important properties. It is generally recognized that the fired color of the clays has been the major cause of differences between the firing technology in the whitewares industry of Europe and Japan and that of the United States.

The fired whiteness of clays is inversely proportional to the amount of iron and titanium oxides that are present. Titanium may be present as rutile or anatase but never combined in the structure of the clay minerals. It is always present in clays mined in the United States as a black amorphous mineral known as leucoxene. Sometimes iron oxide may be present as hematite or one of the other iron oxide minerals, but it too is a constituent of leucoxene. At times iron ions are found in the structure of the clay minerals, particularly montmorillonite (smectite).

Little is known about the origin of leucoxene except that it has generally been accepted as a weathering product of ilmenite; however, there is no explanation for its general deposition between the layers of kaolinite plates. It is fine-grained, amorphous and at times globular in appearance, but the gelatinous material will pass through a 325 mesh sieve opening. The specific gravity is above 4, so when the kaolinite plates are delaminated in water suspension the leucoxene may be removed by centrifuging. The brownish-black color of the leucoxene is attributed to the color of suboxides of both iron and titanium that are present in various ratios. The advent of delamination of the kaolinite stacks with the removal of the leucoxene has allowed the production of a much whiter kaolin product for use in paint, pigments and paper where the raw color of the clay is very important.

The leucoxene is evenly distributed throughout the whiteware body as a thin deposit between the kaolinite plates and so contributes to the general darkening of the ware. Generally, the fired color of the clay may be assessed by obtaining the total iron oxide plus titanium oxide reported in the chemical analysis. However, this is not always an accurate assessment because some of the iron oxide may be present in the clay mineral lattice with varying effects upon fired color.

Fired color assessments are difficult to make because the eye is unable to relate accurately the whiteness of one color tile placed close to another. For the eye to compare whiteness accurately, the specimens must be in the same plane with no line of demarcation between the two. To achieve this state, a simple color control test was devised. Specimens of two plastic clays are placed in the opposite sides of the chamber of a piston extruder so that the resulting extruded piece has one clay along the length on one side with the other clay along the length of the other side. The specimens are fired and very slight variations in whiteness between the two clays may be easily evaluated. The preparation of the specimens may be varied slightly by making plastic tile by hand so that one clay is exposed on one side while the other clay is exposed on the other side. However, in this case a sharp knife cut is required to give the sharp line between the two clays that is required for optical evaluation.

Firing conditions play a very important role in fired color because the suboxides of iron and titanium present in the leucoxene must be oxidized during the preheat period of firing before dehydroxylation of the clays, at which time shrinkage occurs. Shrinkage closes pores in the ware and limits access of oxygen to the interior of the ware. Large amounts of excess air are required in the early stages of firing before the clay shrinkage in order that the leucoxene be oxidized to a creamy or yellowish color. Lack of proper oxidizing kiln atmospheres causes the color of the interior of the ware to be either grayish, bluish or black, and so this condition is called blue coring or black coring, depending upon the color of the core. No adverse physical properties have been found on ware that is black cored; it is merely an indication of improper firing practice. Generally speaking, larger cross sections of ware require longer periods for firing before clay dehydroxylation with larger amounts of excess air in the kiln atmosphere.

The leucoxene found in clays mined in the United States has approximately equal amounts of titanium and iron oxides, while in Europe and Japan, the clays contain a preponderance of iron oxide or titanium oxide. Whereas the leucoxene in domestic clays requires firing in a strong oxidizing atmosphere with large amounts of excess air which results in a creamy color, the leucoxene in European or Japanese clays may be fired under neutral or slightly reducing conditions to give a bluish white color. This color is considered aesthetically more desirable than the yellowish white color of ware in the United States. There are a number of other advantages of the reducing firing techniques used in Europe and Japan over the oxidizing firing techniques used in the United States.

Firing whiteware in a strongly oxidizing atmosphere, required by the equal proportions of iron and titanium oxides in the leucoxene, causes overfiring problems that result in a very narrow firing range for U.S. ware. After the ware matures in a very narrow temperature range close to 1260°C (2300°F) by attaining maximum bulk density, the ware begins to expand or bloat with closed pores causing a sharp decrease in the fired strength. Ware fired in Europe or Japan under neutral or reducing conditions does not overfire but continues to shrink and gain fired strength until ware softening occurs. Thus, this ware is usually fired to a higher temperature, has a desirable bluish white color and has higher fired strength with fewer technological firing problems.

REFERENCES

1. Norton, F. H., *Refractories,* McGraw-Hill Book Co., Inc., N.Y., 1942.
2. Alexander, J., *Colloid Chemistry, Principles and Applications,* D. Van Nostrand Co., Inc., N.Y., 1939.
3. Ormsby, W. C. and Odom, P. R., "Particle Size Analysis of Whiteware Clays: Interlaboratory Comparison of Methods," *J. Mat.,* 5, 3: 586–601 (1970).
4. Binns, C. F. and McMahon, J. F., *The Potter's Craft,* 4th ed., D. Van Nostrand Co., Inc., Princeton, N.J., 1967.
5. Norton, F. H., "Critical Study of the Differential Thermal Analysis Method for the Identification of Clay Minerals," *J. Am. Ceram. Soc.,* 22: 54 (1939).

6. ———, *Elements of Ceramics,* Addison-Wesley Publishing Co., Reading, Mass., 1957.
7. Brindley, G. W. and Nakahira, M., "Kinetics of Dehydroxylation of Kaolinite and Halloysite," *J. Am. Ceram. Soc.,* 40: 346 (1957).
8. ———, "The Kaolinite-Mullite Reaction Series, II, Metakaolin," *J. Am. Ceram. Soc.,* 42: 7 (1959).
9. ———, "A Comparison: Porcelain Insulator Manufacturing Methods in Europe and America," *Ceram. Age,* 25–28 (May, 1969).
10. Gould, R. E., "Electrical Porcelain: Oxidation Vs. Reduction," *Ceram. Age,* 23–26 (February, 1972).
11. ———, "Reduction: A Better Way to Fire Porcelain," *Ceram. Ind.,* 12–15 (August, 1972).
12. Zsolnay, L. M., "Comparison of the American and European High Voltage Porcelain Insulator Manufacturing Technologies and Insulator Design," *Interceram.,* 2: 141–143 (1968).
13. Milliken, W. U., "Coloration of Fused Alumina," B.S. Thesis, Alfred University, 1955.
14. Ehrlich, P., "Phase Ratios and Magnetic Properties in the System Titanium-Oxygen," *Z. Elektrochem.,* 45: 362–70 (1939).

CHAPTER 3

THE CLAY-WATER RELATIONSHIP

Inasmuch as clay and water, when mixed together in varying amounts, form the basic starting material for most ceramic-forming processes, it is desirable to have an understanding of the seemingly strange behavior exhibited by these materials.

An often neglected area has been the role that water plays in the development of the properties of clay-water systems. Although water appears a simple liquid, it is actually very complex, and many properties exhibited by water are still unexplainable. The following concepts are helpful in considering the structure of water.

The Water Molecule[1]

The water molecule consists of an oxygen ion, O^{-2}, with two hydrogen ions, $2H^{+}$, forming H_2O, a neutral molecule. The structure of this molecule is illustrated in Figure 3–1. The two hydrogen ions are not located directly opposite each other; there is an angle of 104°40′ between them. Such an arrangement causes the center of gravity of the negative charge of the O^{-2} ion and the center of gravity of the positive charge from the two H^{+} ions to be separated. Therefore the water molecule has one side that is positively charged and one side that is negatively charged. Such a molecule is called a dipole. The fact that the water molecule is a dipole allows for various types of interactions with other molecules and ions that otherwise would not be possible.[2]

DIPOLE-DIPOLE ATTRACTIONS

Because the water molecule acts as though one side were positive and the other negative and because dissimilar charges attract each

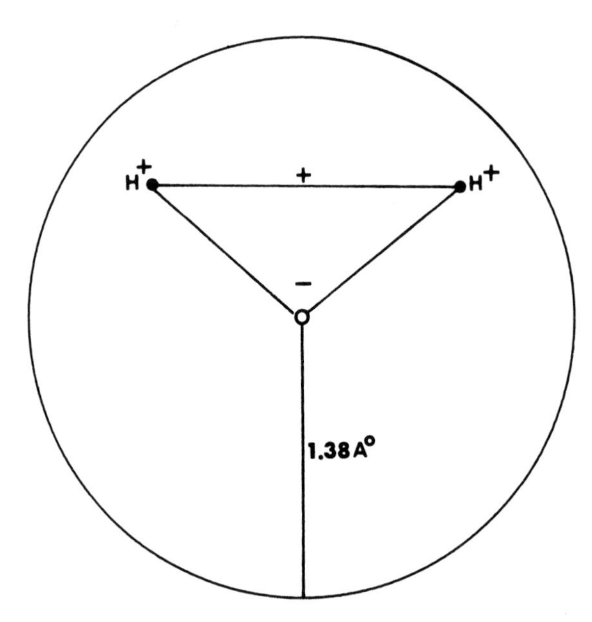

Fig. 3–1 Structure of the water molecule. The H^+ are the two hydrogen nuclei, O the oxygen nucleus; (+) indicates the center of positive charge and (−) the center of negative charge.

other, two water molecules may become associated with each other. Studies have shown that, on the average, each water molecule is surrounded by four others in a tetrahedral arrangement.[3] The positive side of one molecule becomes bonded to the negative side of the adjacent one. This arrangement may be continued, resulting in the hexagonal ring structure shown in Figure 3–2. Such a liquid is an associated or polar liquid.

ION-DIPOLE ATTRACTIONS

It is possible to have attraction not only between water molecules themselves, but also between ions and water molecules. It is known that an ion in water is not a separate entity but is associated with water molecules and travels with a certain number of these neighbors. The number of neighbors depends on the size of the ion and its charge.

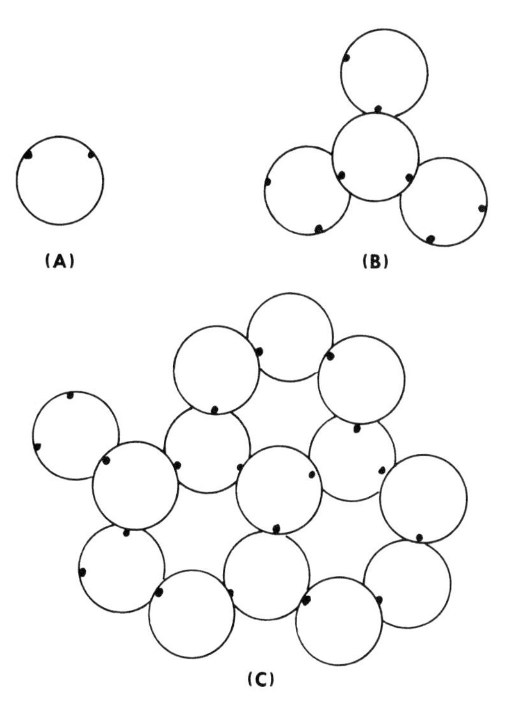

Fig. 3–2 Dipole-dipole attraction resulting in association of water molecules. (A) Single molecule; (B) tetrahedral arrangement; (C) hexagonal ring structure of water.

Figure 3–3 A shows a positively charged ion surrounded by four unoriented water molecules. The arrows indicate the attractive forces between the positive ion and the negative sides of the water molecules. Figure 3–3 B shows the ions associated with the water molecules and in contact with them. They are oriented with the negative side next to the ion and the positive side away from the positive ion.

Effect of Ions on Water Structure

The concepts of dipole-dipole and ion-dipole attractions provide a basis for understanding the effect of ions on water structure. Frank

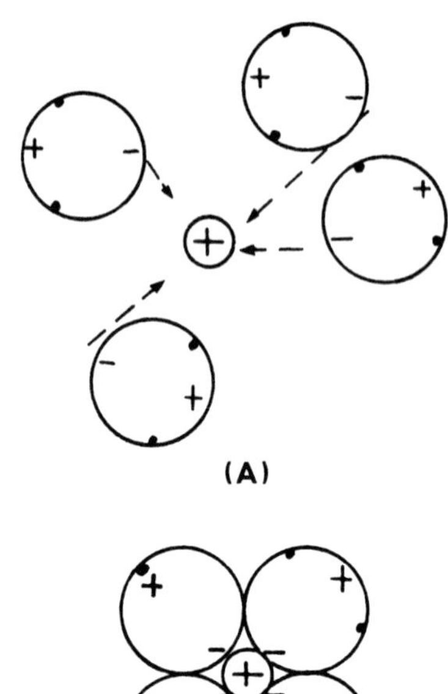

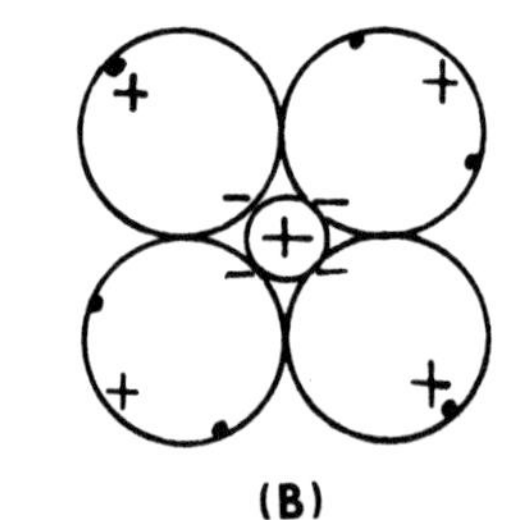

Fig. 3–3 Ion-dipole attraction. (A) Positive ion with four water molecules slightly removed. Arrows indicate attractive forces between positive ion and negative side of water molecules. (B) Water molecules in contact with ion and oriented.

and Wen[4] have presented a simple model, as illustrated in Figure 3–4, which shows the structural modifications in water produced by a small ion.

Surrounding a spherical ion of 2 to 3 angstroms (Å) in radius in a liquid such as water, there is an electric field of 10^6 volts per centimeter (v./cm). Such a strong electric field exerts strong attractive forces and immobilizes the nearest neighbor water molecules as a result of the ion-dipole attraction. This is shown by region A in Figure 3–4. Region C contains normal water having the structure previously described as a result of the dipole-dipole attraction. Region B bridges the gap between the two different types of structure in A and C.

The size of region A can be enlarged or reduced by changing the size or charge of the ion. Small, highly charged ions enlarge the region of immobilization A, while large, singly charged ions reduce it. Numerous investigators have studied the relationship between ion char-

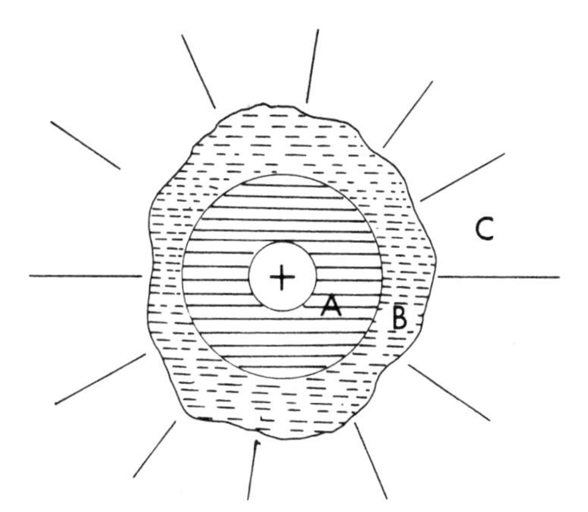

Fig. 3–4 Model for structure modifications in water produced by a small ion. (A) Region of immobilization; (C) region of normal water structure; (B) region of high disorder which bridges the gap between the two different structures A and C.

acteristics and immobilization of water molecules. The data indicate that size is an important factor. Forslind[5] suggests that small ions which fit into the water structure without disrupting it would enhance its development. Larger ions would be expected to retard its development. From the data the critical size for the monovalent cation is approximately 1.36Å radius.[6] This is in close agreement with the size of the "hole" in the hexagonal water structure shown in Figure 3–2. Therefore, it, appears that ions which fit into this hole without disrupting the structure will enhance the build-up of the structure and make it more stable, while larger ions that cannot fit into the hole disrupt the structure. This effect is illustrated in Figures 3–5 and 3–6.

The Water Hull Concept

The idea of immobilization of water induced by ions of high charge and small size provides a basis for the mechanism of formation of a water hull or solvated layer surrounding a clay particle.

Figure 3–7 illustrates in two dimensions the structure of kaolinite. If fracture occurs along the line C-C, bonds are broken, as illustrated in Figure 3–8. The fresh surface formed will have positive and

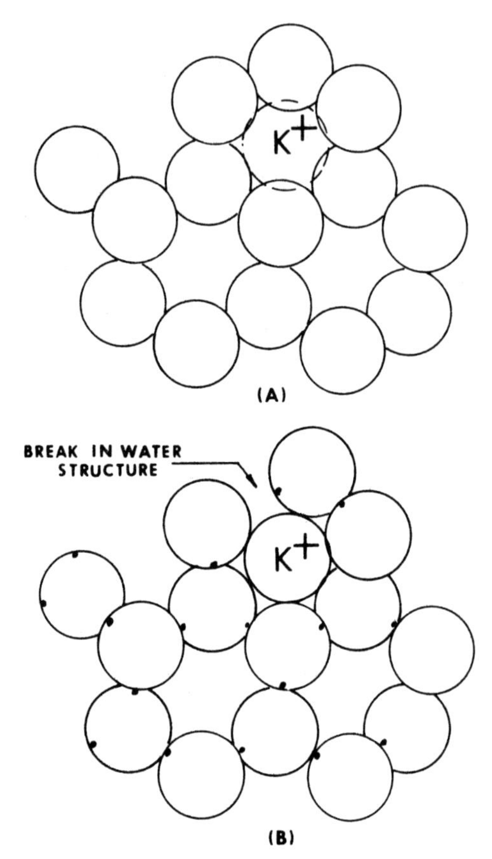

Fig. 3–5 (A) Water structure showing size of K^+ ion with respect to hole in the hexagonal ring. (B) Disruption of continuous water structure when the K^+ ion enters the structure.

negative crystal sites due to the exposure of Al^{+3}, Si^{+4}, O^{-2} and OH^- ions. One would expect such active sites to exert an influence on the surrounding water. This influence might be even more pronounced than that of an individual ion in solution because these ions are on the surface of a rigid crystal lattice. They will have the capability of attracting the dipole water molecule or other positive or negative ions in order to satisfy their charge deficiency.

The extent or thickness of the water layer built up around the particle will depend on the types of ions associated with the water. Ions that promote the structure of water, such as Ca^{+2}, Mg^{+2} and Al^{+3}, will cause a large solvated layer to be built up. The large ions K^+, Rb^+, Cs^+, NH_4^+ and Na^+ will disrupt the water structure and reduce its size.

The importance of the water structure in clay-water systems will

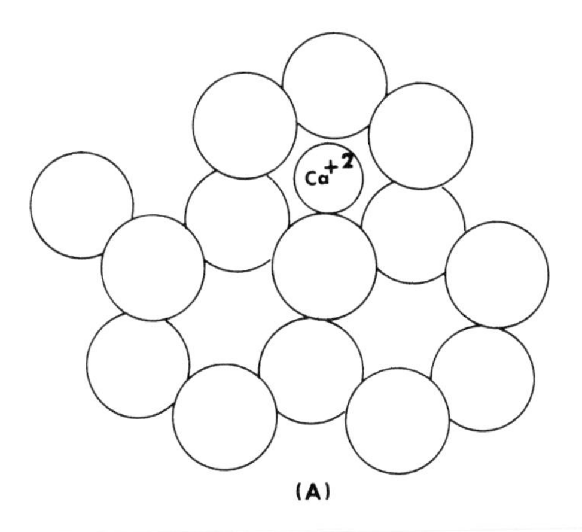

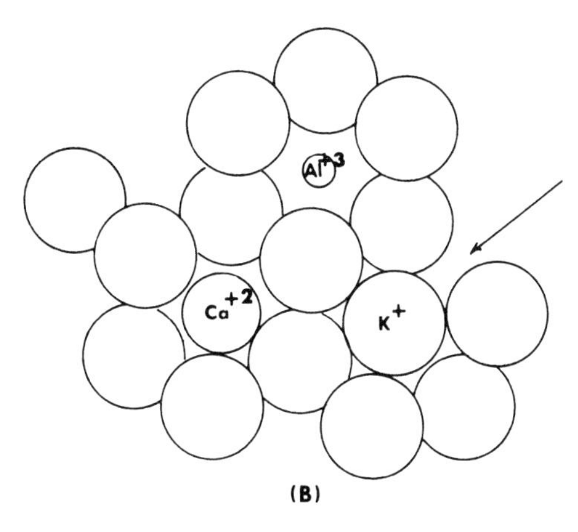

Fig. 3–6 (A) Accommodation of Ca^{+2} ion in water structure. (B) Ca^{+2}, Al^{+3} and K^+ ions and their effect on the continuity of the water structure. Arrow indicates break in structure caused by K^+ ion.

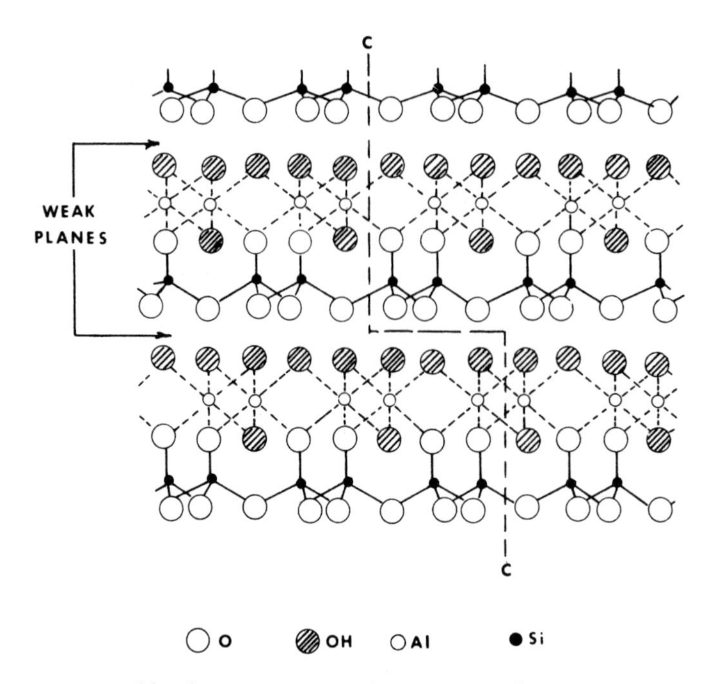

Fig. 3–7 Structure of kaolinite. Fracture along C-C produces new active surface.

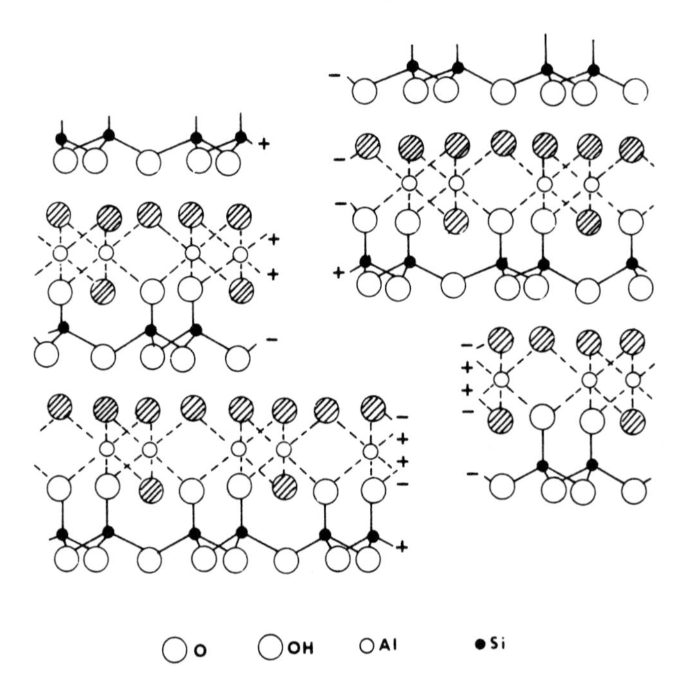

Fig. 3–8 Crystal of kaolinite after fracture along C-C, showing formation of equal numbers of positive and negative sites.

become more obvious when such properties as plasticity and flow are discussed. In systems having small particle size and large surface area as well as high concentration of solids, the separation distance between particles may approach the water film thickness. The properties of this water film exert a strong influence on the shear and flow properties of the clay-water system.

The Relationship Between Kaolinite Crystal Lattice and Water Structure

To complete the picture of the adsorption of water molecules by the surface of clay minerals, a discussion of the relationship between the structures of the two phases is necessary. If the geometric positioning of the ions and molecules is such that there is a good fit and no large

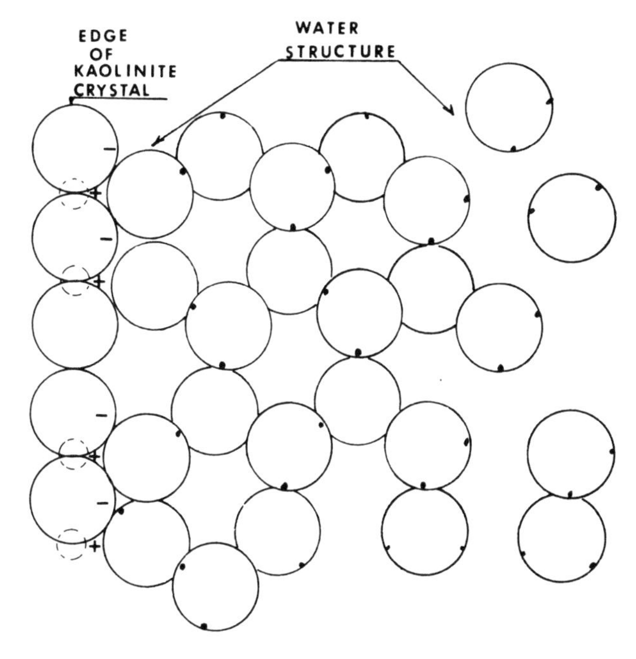

Fig. 3–9 Kaolinite surface (left) and build-up of hexagonal water structure on this surface.

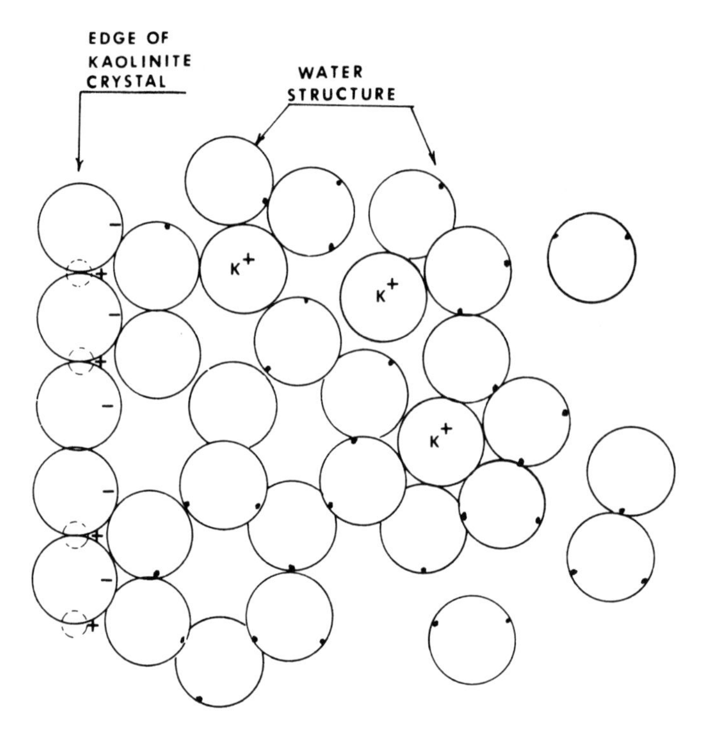

Fig. 3–10 Kaolinite surface and water structure resulting when K^+ ion is present.

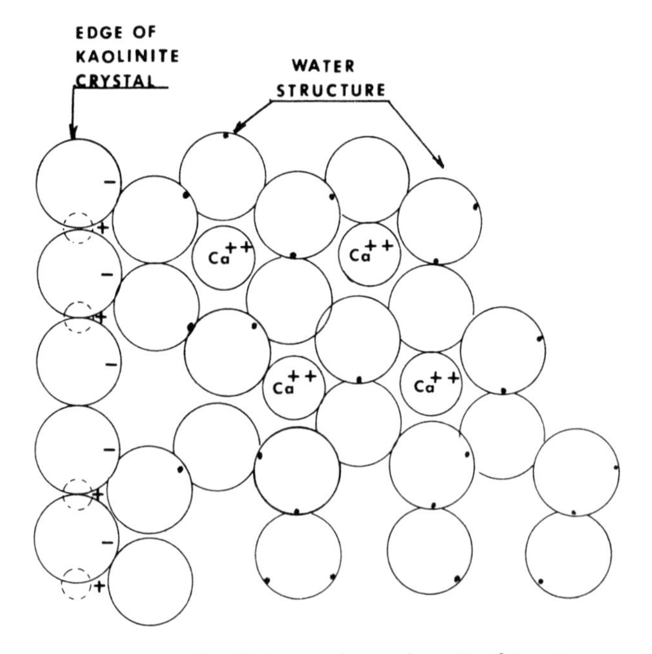

Fig. 3–11 Water structure on kaolinite surface when Ca^{+2} ion is present.

degree of disorder at the interface, one would expect good bonding and adherence between the two. As previously indicated, water structure is a hexagonal ring arrangement with each water molecule tetrahedrally surrounded by four neighbors. This structure has the same geometric dimensions as the layer silicates.[7] The degree of misfit between the kaolinite crystal and the water structure is actually only -1.1 percent.[8] This fact probably is the main reason why clay-water systems exhibit their unusual and useful properties.

Figure 3–9 illustrates the build-up of the water structure on the edge of a kaolinite crystal indicating the good fit between the two structures and the hexagonal ring arrangement of the water molecules. Figure 3–10 shows the effect of the large K^+ ion on the water structure. It is obvious that the addition of such an ion disrupts the structure and would inhibit the build-up of any structure to a great distance. It is also obvious that it would take very little force to shear such a system.

Figure 3–11 shows the structure with the Ca^{+2} ion fitting into the interstices of the water structure. The incorporation of such an ion in this position would greatly strengthen the structure and enhance its build-up and extension to large distances from the surface of the crystal.

These concepts provide the basis for understanding the behavior of casting slips as well as plastic bodies. The large, singly charged "structure-breaking ions" are used as deflocculants in casting slips because low viscosity, high solids content, low or zero yield point are desired. On the other hand, smaller, multicharged ions can be used to improve the plasticity of clay bodies when a high yield point is desired.

REFERENCES

1. Bernal, J. D. and Fowler, R. H., *J. Chem. Phys.*, 1: 515 (1933).
2. Moeller, T., *Inorganic Chemistry*, p. 187, John Wiley & Sons, Inc., N.Y., 1952.

3. Morgan, J. and Warren, B. E., *J. Chem. Phys.*, 6: 666 (1938).
4. Frank, H. S. and Wen-Yong, W., "Interactions in Ionic Solutions," *Diss. Faraday Soc.*, 24: 133–41 (1957).
5. Forslind, E., "A Theory of Water," Swedish Cement and Concrete Research Institute at the Royal Institute of Technology, Stockholm, 1952.
6. Lawrence, W. G., ed., *Clay-Water Systems*, SUNY College of Ceramics, Alfred, N.Y., 1965.
7. Hendricks, G. B. and Jefferson, M. T., "Structure of Kaolin and Talc-Pyrophyllite Hydrates and Their Bearing on Water Sorption of Clays," *Am. Mineral.*, 23: 863–75 (1938).
8. Mason, B. J., "The Growth of Snow Crystals," *Sci. Am.*, 204: 1 (1961).

C H A P T E R 4

CLAY SLIPS

The ability of a clay to be dispersed in water by the addition of certain chemicals known as deflocculants serves as a base for the preparation of casting slips in which clay and nonplastic ingredients, such as flint and feldspar, compose the mixture. A deflocculated slip provides a stable suspension having minimum viscosity consistent with the desirable high solids content. It is the necessary starting point for all slip-casting operations.

Development of Charge on Clay Particles

The manner in which deflocculants work is the basis for understanding the behavior of casting slips.

As stated previously, the kaolinite crystal has unsatisfied sites on its surface. These (+) and (−) sites have the ability to attract ions or molecules of opposite charges (Figure 4–1).

If a kaolinite crystal is placed in water containing sodium hydroxide, NaOH, the Na^{+} ion will be adsorbed at the (−) sites and the OH^{-} at the (+) sites. Because the crystal surface has an equal number of positive and negative sites, one would expect the particle to remain neutral (have no overall charge) if each site adsorbed an ion of opposite charge. However, this is not the case. Clay particles that seem to have a full complement of adsorbed ions actually exhibit a negative charge.

The energy with which an ion is bonded to the surface of a crystal will depend on the size and charge on the ion. Ions that have a small charge will be held less rigidly than those having a high charge. Small ions will be held to the surface of the crystal more tightly than large ions. Thermal energy maintains ions in a constant state of motion.

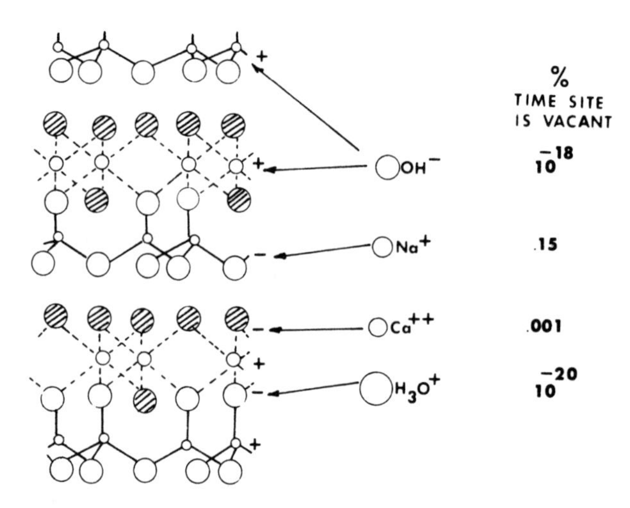

Fig. 4–1 Adsorption of (+) and (−) ions on the edge of crystal sites of kaolinite. Numbers indicate percentage of time the site is vacated by the adsorbed ion as a result of thermal energy in the system.

If the thermal energy is sufficient, it may break the bond between the ion and the crystal surface and that ion may wander back into the solution. Thus, there is a discontinuous bonding of the ions to the crystal surface; the smaller the charge and the larger the size of the ion, the greater the discontinuity of bonding. The percentage of the time an ion is bonded with the crystal surface may be calculated.[1] In the case of the Na^+ ion, calculation indicates that it leaves its position on the crystal surface 0.15 percent of the time. On the other hand, the OH^- ion stays on the crystal surface a greater percentage of the time, leaving its position only 10^{-18} percent of the time. In other words, the negative crystal sites are vacant for longer intervals than the positive sites, and the crystal under these conditions has a net negative charge. The low-bond energies of the large monovalent cations, such as K^+, Na^+, Cs^+ and $NH_4{}^+$, result in the development of a negative charge on the kaolinite particle. The smaller highly charged cations, such as Ca^{+2}, Mg^{+2}, Al^{+3}, are more strongly held in their adsorbed positions and do not wander away once they are adsorbed. Under these conditions, the particle cannot develop a charge.

If all of the particles in a clay slip are negatively charged, they will repel each other, remain separated and stay in suspension longer. The resulting slip is stable. Blunging (mixing with water) breaks down agglomerates of particles into individual particles, exposing fresh surfaces, and allows the ion exchange mechanism previously described to become complete and uniform throughout the slip. This process takes time; hence, the viscosity of a slip may change in the first several hours, and proper blunging and aging are desirable.

This is one of several theories proposed to explain how the charge is developed on clay particles. Evidence supporting this theory is as follows:

1. The calculated charge on a particle based on this theory closely coincides with the observed charge.
2. Increase in temperature should and does cause an increase in charge on the particle. This is confirmed by Button.[2]
3. Theory predicts that a positive charge should be developed on a kaolinite particle if a small highly charged cation, such as Al^{+3}, and a large singly charged anion, Cl^- or CNS^-, were adsorbed. This has also been confirmed by Button.[2]

Ions that are capable of producing a charge on a clay particle disrupt water structure, whereas those that produce little or no charge promote water structure. These two effects provide the basis for flocculation and deflocculation.

Changes in Viscosity

One of the properties most sensitive to the presence of monovalent cations, such as Na^+, is viscosity.[3] Figure 4–2 shows the viscosity of a kaolinite suspension as NaOH is added. It is seen that the viscosity remains at the original high level as the first additions of NaOH are made and that at a certain point there is a sudden decrease in viscosity to a small fraction of the original value.

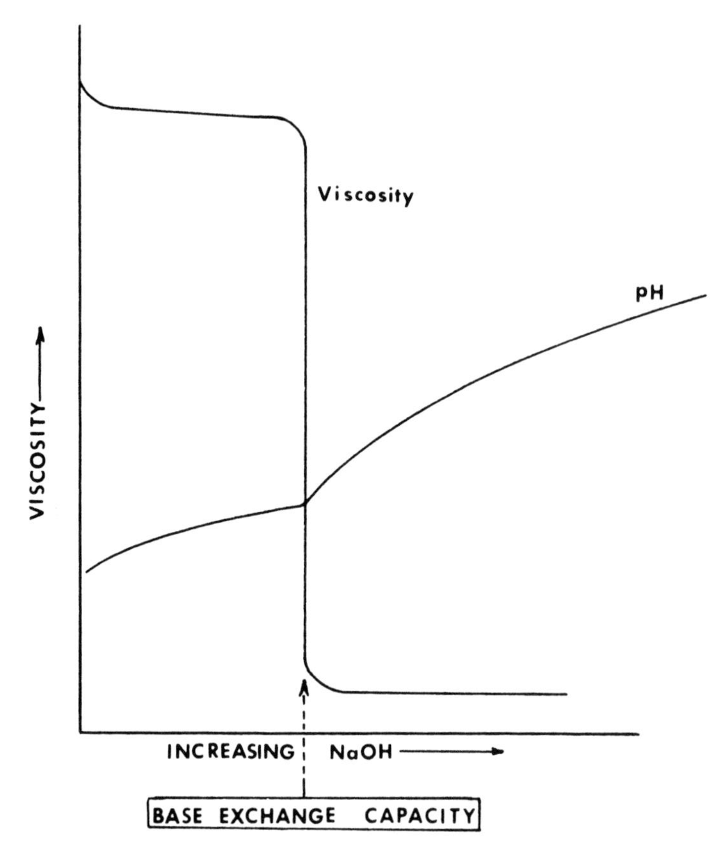

Fig. 4–2 Viscosity and pH of a pure kaolinite suspension versus NaOH addition.

The hydrogen ion concentration (pH) indicates the concentration of H_3O^+ and OH^- ions in a solution. An acid solution has a low pH and a basic solution a high pH. The pH scale is 1 to 14 with a pH of 7 being neutral. The pH of the pure kaolinite suspension shown in Figure 4–2 is 5.5. As NaOH is added, the pH rises very slightly until the same point is reached at which the viscosity drops. At this point there is a sudden rise in pH indicating that the concentration of OH^- ions is increasing rapidly in the solution.

As NaOH is initially added to the suspension of kaolinite and water, Na^+ and OH^- ions are formed. The Na^+ ions will be attracted by the (−) crystal sites and the OH^- ions by the (+) crystal sites. Up to the point where the (+) and (−) sites are filled, those ions are removed from the solution and there is very little change in viscosity or pH. However, when all the crystal sites are filled, further additions cause the viscosity to drop and the pH to rise. This amount of added electrolyte is known as the exchange capacity of the clay. Beyond this amount the Na^+ and OH^- ions are no longer removed from the solution. The Na^+ ions disrupt the water structure, the rigid water layer is no longer as thick and a dramatic drop in viscosity is observed. At the same time the OH^- ions cause the rise in pH.

The exchange capacity depends on the particle size distribution of the clay. This is to be expected because small particle sizes will have greater surface area and more crystal sites to be filled by the adsorbed ions.

Although NaOH has been discussed as a deflocculant for pure kaolinite systems, it seldom works in practice. This is due to the fact that most clays have small amounts of soluble salts containing calcium. As indicated previously, the Ca^{+2} ion is held by the kaolinite crystal much more strongly than the Na^+ ion. To replace the Ca^{+2} ion with Na^+, the Ca^{+2} must be removed from the system by a chemical reaction that will tie the calcium ion up in an insoluble condition. Figure 4–3 shows the effectiveness of three different sodium salts as deflocculants for a clay that is contaminated with calcium. In the case of a NaOH addition, the Ca^{+2} ion would react with the OH^- ion to form calcium hydroxide: $Ca^{+2} + 2(OH)^- \rightarrow Ca(OH)_2$. Calcium hydroxide is slightly soluble and, therefore, Ca^{+2} and OH^- ions are again formed in the solution. The Ca^{+2} has not been removed from the system. In this case it is necessary to deflocculate with a sodium salt containing an anion that will react with the calcium to form an insoluble calcium salt.[4] Sodium silicate, Na_2SiO_3, works well in this case. The Ca^{+2} ion reacts with the $SiO_3{}^{-2}$ ion to form insoluble $CaSiO_3$. In practice, mixtures of sodium carbonate, Na_2CO_3, and sodium silicate, $2Na_20 \cdot SiO_2$, sodium phosphate, Na_2HPO_4, or organic deflocculants are usually used.

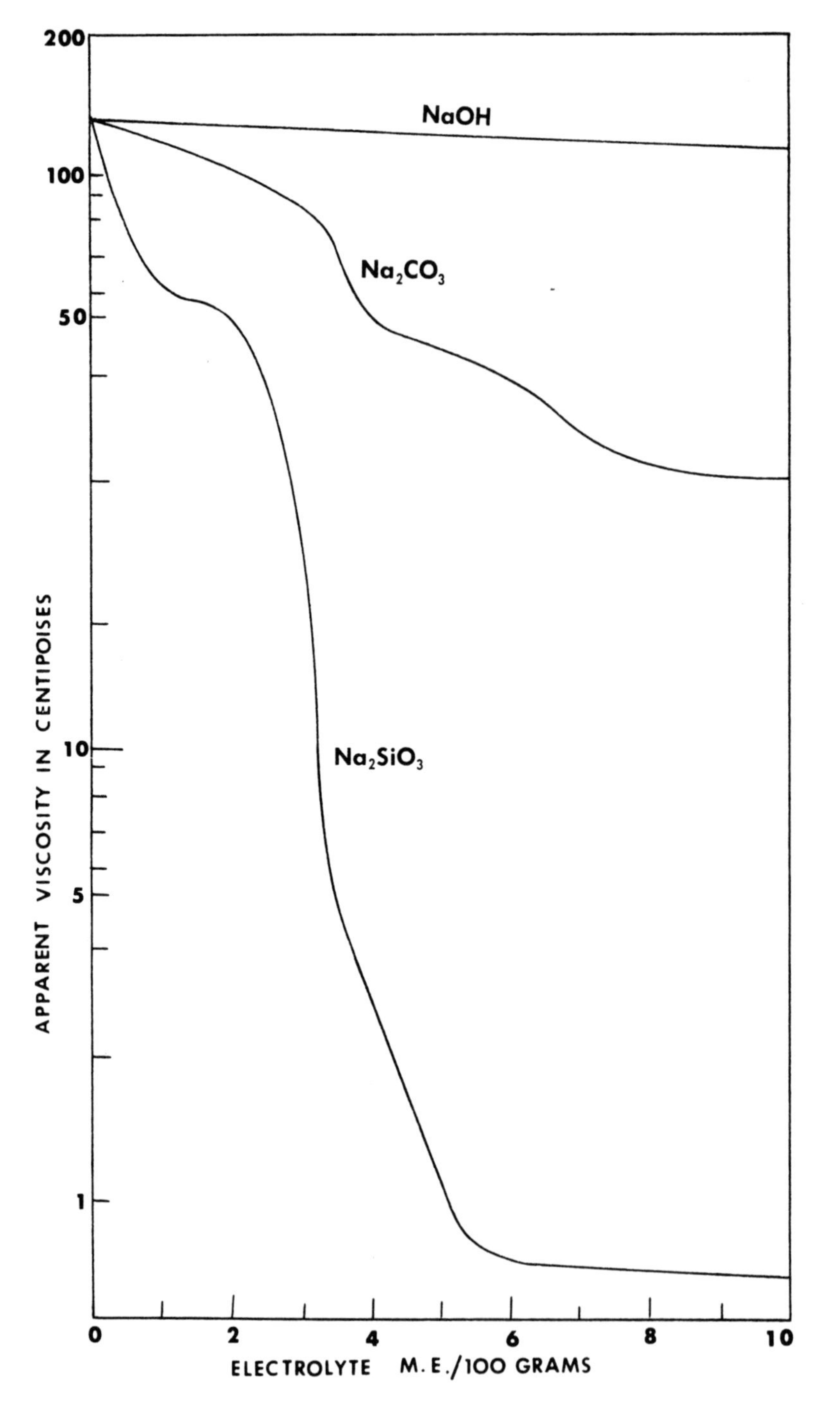

Fig. 4–3 Viscosity of kaolinite suspension purposely contaminated with Ca^{+2} ions. Effectiveness of the three different sodium salts as deflocculants is illustrated (after Johnson and Norton).

Soluble Sulfates in Slip Control

Many clays contain small amounts of gypsum, $CaSO_4{\cdot}2H_2O$. The calcium ion has a strong flocculating effect that causes the clay to be very workable, dry rapidly and also have a higher water of plasticity when compared to clays containing alkali ions. Precipitation of the calcium ion or removing it from contact with the surface of the clay may cause a marked decrease in the workability of the clay-water paste.

The sulfate ion may migrate to the surface of the clay ware during drying causing an unsightly whitish deposit known as drier scum, or it may decompose underneath the surface of a glazed article during firing to cause bubbles in the glaze.

The sulfate ion content is easily measured in a clay-water slip[5] and is one control test closely watched for proper casting or plastic forming. A range of sulfate ion in the clay-water slip from 100 to 500 parts per million (ppm) is usually maintained, as readings above or below the desired range are a fairly reliable indicator of forming, drying and firing problems.

The role of the sulfate ion upon the clay is still quite contentious, but there is general agreement upon the methods for handling the ion, whether dealing with slip casting, drier scum or workability of clay-water pastes.

Table 4–1 gives data[6] on the solubility of calcium and barium compounds in water. This information provides the basis for the control of the various defects caused by the presence of $CaSO_4{\cdot}2H_2O$ in clays.

The control over the soluble sulfate concentration in whiteware slip has been regarded as a primary control property for the successful slip casting or plastic forming of ware. The actual function of the sulfate ion has been debated with some people reporting that it causes mild deflocculation and others contending that it causes flocculation. Although the function of cation adsorption by the clay seems clear, the size and bond strength of the chloride and sulfate ions seem to

TABLE 4–1
SOLUBILITY OF CALCIUM AND BARIUM COMPOUNDS IN WATER[6]

		WATER SOLUBILITY			
		Cold		**Hot**	
Compound	**Mineral**	**Gm./100 ml**	**T°C**	**Gm./100 ml.**	**T°C**
$BaCO_3$	Witherite	0.0022	18	0.0065	100
$BaCl_2$		31.0	0	59.0	100
$BaSO_4$		0.00022	18	0.00041	100
$Ba(OH)_2$	Barite	5.6	15	94.7	78
$CaSO_4 \cdot 2H_2O$	Gypsum	0.241	0	0.222	100
$CaSO_4$	Anhydrite	0.209	30	0.1619	100
$CaCO_3$	Calcite	0.0014	25	0.0018	75
$CaCl_2$		59.5	0	159.0	100

negate their adsorption by the clay. There is little evidence to indicate that chloride ions are adsorbed by the clay but the variations in soluble sulfate in slip seems related to variations in the cations—particularly calcium. In all probability the sulfate ion plays a passive role in adsorption but has an important function in the controlling of the adsorption of cations.

Gypsum dissolves with difficulty in water to give a saturated solution containing the following proportion of ions:

Gypsum solubility: 0.2 g./100 ml.
2.0 g./L.
2000 ppm

Calcium ion in saturated solution:
$2000 \times 40/136 = 588$ ppm

Sulfate ions in saturated solution:
$2000 \times 96/136 = 1412$ ppm

Rarely would the slip reach a saturated level of solution because usually this much gypsum is not present in the clay. Some of the calcium ions are attracted to the surface of the clay particles causing them to have flocculated properties. Filter pressing the clay-water slip removes most of the soluble sulfate ions, but many calcium ions remain on the surface of the clay particles.[7,8] Reusing the filtrate causes a build-up of soluble sulfate above 1412 ppm which represents their amount in a saturated solution. This build-up causes a precipitation of calcium ions[9] in the slip, reducing the number of calcium ions available for flocculation of the clay.

Let us assume that the soluble sulfate ions increase to 1600 ppm or 1600 g./1000 L. The concentration of the soluble sulfate ions is (1600/96)/1000 or 0.01667 moles (m.)/L. Slightly soluble compounds have what is known as a *solubility product* and so calcium sulfate precipitates when the product of the concentrations of calcium ions times the concentration of the sulfate ions is greater than 1.95×10^{-4}, which is the solubility product for calcium sulfate:

$$C_{Ca^{+2}} \times C_{SO4^{-2}} = 1.95 \times 10^{-4}$$

When the concentration of the sulfate ions reaches 0.01667 m./L. (1600 ppm), all calcium ions over the following amount will precipitate as calcium sulfate:

$$C_{Ca^{+2}} \times 0.01667 = 1.95 \times 10^{-4}$$

or

$$C_{Ca^{+2}} = (1.95 \times 10^{-4}/0.01667 = 0.0117 \text{ m./L.}$$

or

$$C_{Ca^{+2}} = 0.0117 \times 40 \times 1000 = 468 \text{ ppm}$$

Thus when the level of soluble sulfate in the slip increases to a level of 1412 ppm or higher, the amount of soluble calcium available for flocculation of the clay is decreased.[8] This is a condition probably attained frequently in localized areas in clay-water slips.

Barium carbonate is not a suitable additive for reducing the soluble sulfate in whiteware slip because it is so slightly soluble and it reduces the calcium ions. Barium chloride or barium hydroxide are more suitable because they reduce the sulfate selectively without interfering with the calcium ions. They are both quite soluble in a water solution. The procedure for determining the proper amount of addition will be discussed in Chapter 6 in the section on drier scum.

At some times the soluble sulfate content of the slip becomes too low, such as below 100 ppm. The sulfate content has little effect upon the slip but is an indication that little gypsum has been furnished by the clay and so there are insufficient calcium ions to produce the flocculation level normally desired in the slip. To correct this condition, a slight amount of pottery plaster, $CaSO_4 \cdot \frac{1}{2}H_2O$, is added to the slip.[9]

Summary

The following list shows some of the main differences between flocculated and deflocculated clay-water systems.

Flocculated	*Deflocculated*
Rapid settling of suspensions	Stable suspensions
Low solids content	High solids content
High viscosity	Low viscosity
High yield point	Low yield point; slumps under own weight
Higher shrinkage	Lower shrinkage
Lower green bulk density	High green bulk density
Lower green strength	High green strength

Many other factors are involved in making a good casting slip. The starting materials, soluble salt content of the materials, method of slip preparation, control of deflocculation and control of specific gravity of the slip all play important roles. Norton[4] has given an excellent treatment of this subject, and it is suggested for further reading.

REFERENCES

1. Lawrence, W. G., "Theory of Ion Exchange and Development of Charge in Kaolinite-Water Systems," *J. Am. Ceram. Soc.*, 41: 4 (1958).
2. Button, D. D. and Lawrence, W. G., "Effect of Temperature on the Charge on Kaolinite Particles in Water," *J. Am. Ceram. Soc.*, 47, 10: 503–509 (1964).
3. Johnson, A. L. and Norton, F. H., "Fundamental Study of Clay, II, Mechanism of Deflocculation in the Clay-Water System," *J. Am. Ceram. Soc.*, 24, 6: 189–203 (1941).
4. Norton, F. H., "Ceramic Forming with Casting Slips," Ch. 9, *Fine Ceramics, Technology and Applications*, McGraw-Hill Book Co., Inc. N.Y., 1970.
5. ———, Standard Test Method for Soluble Sulfate in Ceramic Whiteware Clays (Photometric Method), Am. Soc. for Test. and Mat., Designation C 867–77, 1979.
6. Handbook of Chemistry and Physics, Chemical Rubber Publishing Co., Cleveland, Ohio.
7. West, R., "Characteristics of Filter Pressed Kaolinite-Water Pastes," Clays and Clay Mineral Conference, Proc., 12th Nat. Conf., MacMillan, 1964.
8. Shell, H. R. and Cortelyou, W. P., "Soluble Sulfate Content of Pottery Bodies During Preparation, *J. Am. Ceram. Soc.*, 26, 6: 179–85 (1943).
9. Johnson, P., "A Proposed Mechanism for the Release of Sulfate Ion Into the Filtrate of Georgia No. 600 Kaolin," B.S. Thesis, Alfred University, 1971.

CHAPTER 5

PLASTIC PROPERTIES

Plasticity may be defined as the property of a material that permits deformation when an external force is applied and retention of the deformed shape when the force is removed. It may be defined more quantitatively as the product of yield point and extensibility. The yield point is the force required to start movement or to cause the initial deformation. Extensibility is the amount of movement that the system can tolerate before cracks appear or fracture results.

A system may have a very suitable yield point but little extensibility. This type of clay would be described as being short.

Other clays may have both a suitable yield point and extensibility but a very narrow working range. That is, the clay is very sensitive to water content; with a very slight addition of water above the optimum it becomes sticky. Numerous intangible properties, such as stickiness, butter quality, self-lubrication and dryness, are produced by deformation, which add confusion to the interpretation of plasticity measurements.

Various types of elaborate mechanical testing equipment have been built to study and measure the plastic properties of clays, and much has been learned as a result of these studies.

One of the qualitative methods of measuring the plastic properties of a clay-water system has been described by West.[1] This involves the use of the Braebender Plastograph, a mechanism that continuously records the force required to shear a clay-water system as water is added at a controlled rate. Figure 5–1 shows a typical curve generated during such a test and gives a picture of the plastic properties of the clay as a function of water content.

One starts the test with a known amount of dry clay. The particles are in contact or separated by air films between particles. The

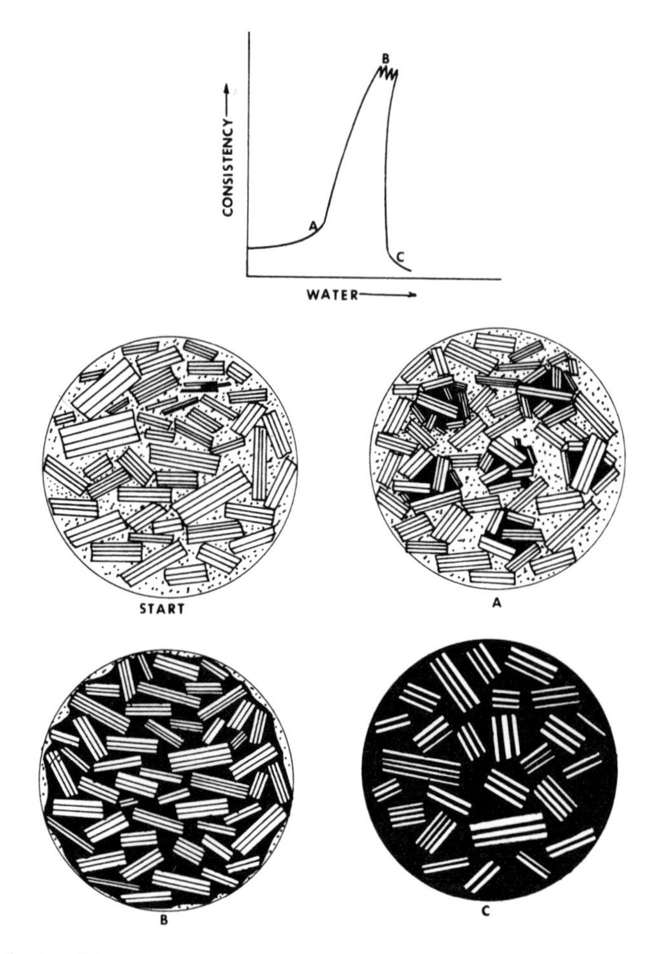

Fig. 5–1 Relationship between consistency of a clay-water system and water content. Inserts (below) represent relationship of clay-air-water as related to the consistency curve above.

clay behaves as a dry powder and requires little force to shear or move one particle with respect to another. When the first drop of water is added to the dry clay, the particles in the immediate area of the droplet will be pulled together into a small spherical granule. Each granule is held together by the surface tension of the water. As more and more of these granules are formed, the shearing mechanism, having a close clearance, begins to break these granules down into smaller ones. This causes a very slight increase from the zero reading, but the consistency remains poor over a wide range of water addition. The individual granules are still separated by dry powder and the forces required to shear the system are low. Because of the surface tension of the water, the forces of cohesion within the granule are much greater than any adhesion between granules.

At point A in Figure 5–1, when all of the clay is in the form of small individual balls or granules and there is no loose powder between granules, the next drop of water causes the granules to agglomerate or stick together. As soon as the necessary water is added to cause adhesion between granules, the force required to shear the system increases rapidly. With very little water beyond point A, point B is reached, which is the maximum in the consistency curve. This point indicates the amount of water required to obtain the maximum plasticity and is known as water of plasticity. For example, a ball clay might have 21.7 percent water at point A, 28.8 percent at B and 42.5 percent at point C. At point B the pores are completely filled with water, there is no longer any air in the system and the particles are separated slightly by water films. The character of these water films determines to a large extent the force required to shear the system.

The range of water content over which a clay maintains a high degree of plasticity is important because this is an indication of the tolerance the clay has for water. (In a throwing operation, it is desirable to have a clay that maintains its high yield point over a wide range of water content. Otherwise, with repeated sponging and wetting, it would slump.) Usually the fine-grained kaolins or the ball clays that may contain some montmorillonite are most tolerant, while the coarse-grained kaolins have a short range of water content.

Beyond the point where maximum plasticity is observed the addition of more water separates the particles, increases the thickness of the water films separating particles and reduces the force required to shear the system. This is illustrated by point C in Figure 5–1.

Factors Influencing Plasticity[2]

The most important factors influencing the plasticity of a given clay-water system are as follows:

1. Type and amount of adsorbed ions that determine the charge on the clay particles.
2. Surface tension of the water.
3. Temperature.
4. Particle size.
5. Rigidity of the adsorbed water film between particles.

In any plastic system, the forces influencing cohesion or attraction between particles must be greater than any repulsive forces. Any factor that reduces these forces will decrease plasticity.

Effect of Adsorbed Ions

It is known that the clay-water systems in which little or no charge is developed on the clay particles show the highest yield point and best plastic properties. Clays having adsorbed monovalent ions that result in charge development or deflocculation have a low or no yield point. With no yield point there is no plasticity.

Other factors being equal, the yield point of a clay decreases as the charge on the particles increases. Plastic properties improve as one changes the type of adsorbed ion from the large monovalent ions, such as Na^{+} and K^{+} to Mg^{+2}, and Ca^{+2} to Al^{+3}. This effect has been discussed in Chapter 4.

The effect of charge is shown in Figure 5–2. A clay containing small amounts of adsorbed Ca^{+2} ions is compared with the same clay having a 1 percent addition of sodium metaphosphate, $NaPO_3$. Ions such as Ca^{+2} or Al^{+3} increase the plasticity of clay-water systems. In the soft-mud brick industry, such additions are used to stiffen the clay when the mined clay is too wet in the spring of the year. A small

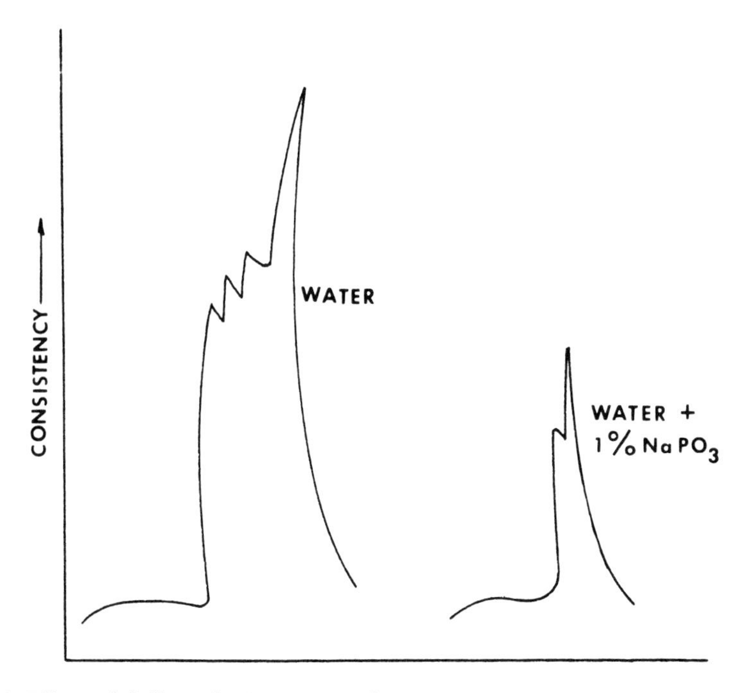

Fig. 5–2 Effect of deflocculation on consistency.

addition of $Ca(OH)_2$ immediately stiffens the clay-water mix to a consistency suitable for the forming operation.

Unfortunately it is impossible to separate the effect of charge on the particle from the effect on the adsorbed water structure produced by these ions. Undoubtedly the latter effect is greater.

Surface Tension of Water

Any clay-water system having a water content in the plastic range must be affected by the surface tension of water. These effects, explained below, have been discussed in detail by Norton[3] in his "stretched-membrane" theory of plasticity.

A section of a clay-water mass near the surface is illustrated in

Figure 5–3. The particles are separated by thin water films that are continuous on the surface and between particles. This is confirmed by the observation that such a mass loses water at the same rate as a free water surface[4] and the fact that shrinkage occurs when a clay-water mass is dried. The thickness of the water film between particles will depend on the water content of the system as well as on the type of ions associated with the system.

In a pure clay-water system, one can calculate the force exerted on particles tending to hold them together. These forces for various sizes of clay particles have been calculated by Norton[3] and measured by Westman.[5] The actual measurements indicate that in the case of a kaolinite the force amounts to 263 psi, while in a finer grained ball clay it is 880 psi. In a stable system, the forces drawing particles together must be opposed by equal forces. Otherwise the system would collapse, the particles would be in contact and a clay-water system would exhibit no plastic properties. The opposing forces involve the charge on the particles that causes them to repel each other and the development of a water structure surrounding the particles that is sufficiently strong to resist the close approach of particles. Because the capillary pressure is inversely proportional to the capillary diameter, this pressure will increase as the particle size of the clay decreases, as confirmed by Norton.[3]

Further work by Kingery and Francl[6] has shown the effect of surface tension of the water phase on the drying and plastic properties. As the surface tension of the water phase is decreased by the addition of surface tension reducing agents, the plasticity and workability are

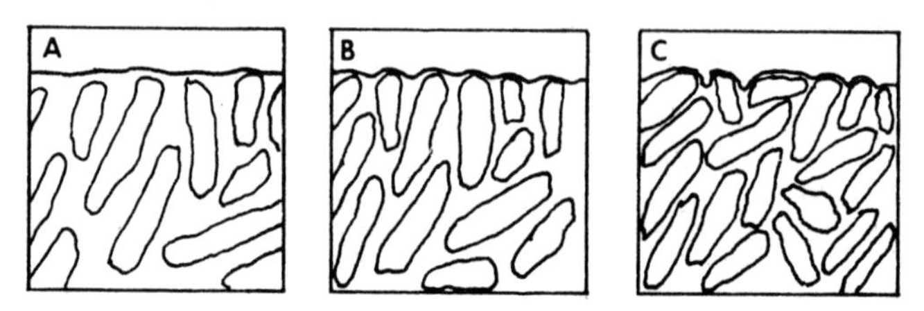

Fig. 5–3 Plastic clay mass. (A) Soft, (B) medium, (C) stiff.

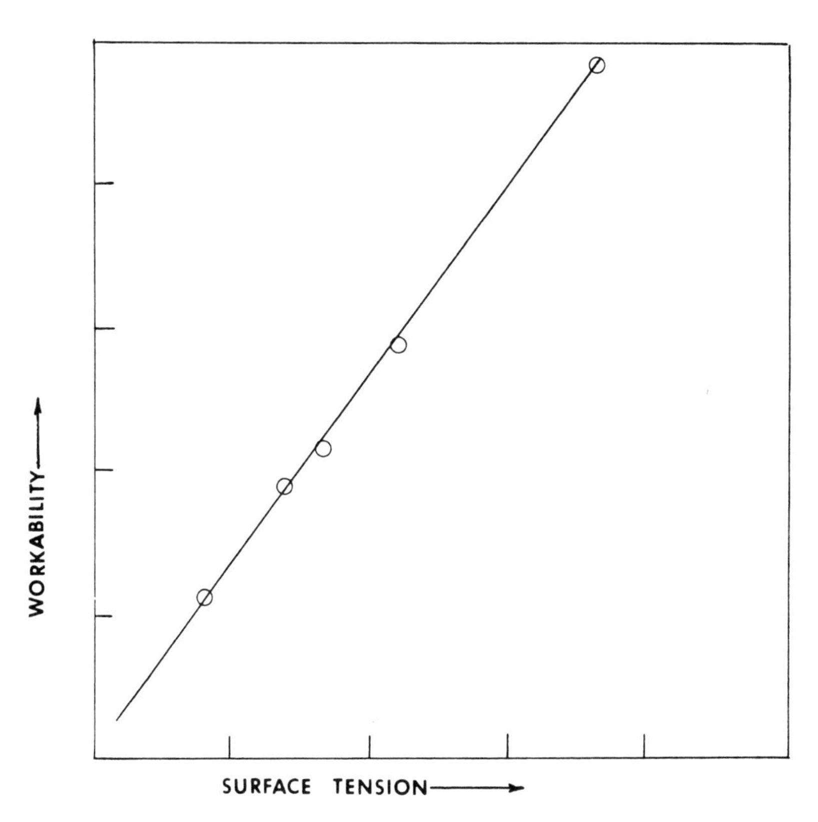

Fig. 5–4 Effect of reduction of surface tension of the water on the workability of a clay-water system.[6]

decreased, as shown in Figure 5–4. This curve indicates that clay would have no workability in a liquid having no surface tension. The effect of surface tension reducing agents on the rigidity of the water structure is not presently known, and this is undoubtedly also involved.

Temperature

The effects of increasing temperature in a clay-water system are as follows:

1. Reduction in rigidity of the adsorbed water films resulting in lower yield points and a decrease in the force necessary to deform the system.
2. Increase in the charge on the particles which, in turn, will lower the yield point of the system.
3. Decrease in the surface tension of the water which will lower the yield point.

Because of these temperature effects, we have seen in recent years the development of hot mixing and hot plastic extrusion of clay-water systems. Because the system can be deformed more easily at the higher temperature, less water can be used, resulting in less shrinkage during the drying process. The other obvious advantage is the elimination of the heat requirement to raise the temperature of the piece from room temperature to drying temperature. Less temperature gradient is developed in the piece during the drying process.

The effect of increased temperature on the consistency of a plastic clay is shown in Figure 5–5.

Particle Size

Because the capillary pressure tending to hold particles together in a clay-water system is related to the size of the capillaries and the size of the capillaries is determined by the particle size of the clay, one would expect an increase in yield point and plastic properties as particle size is reduced. It is well known that the addition of small amounts of small-particle-size clay minerals, such as montmorillonite (bentonite), to the coarse kaolin clays greatly improves their plastic properties (Figure 5–6).

Another effect of particle size involves the number of water films per unit-distance. With a large number of small particles there will be a greater number of water films per unit-distance than with a small number of large particles.

The structure of water and the effect of ions on that structure

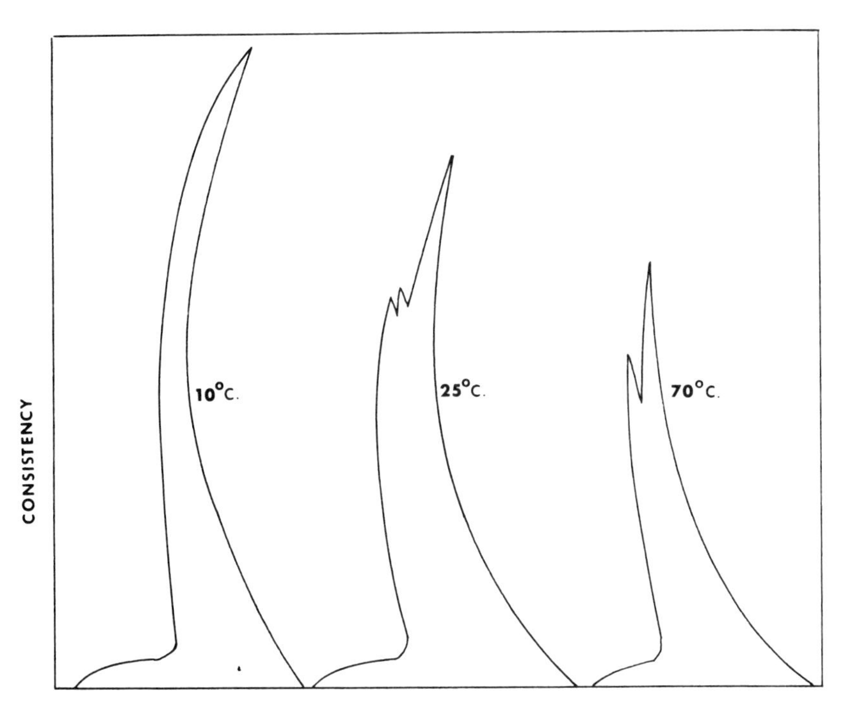

Fig. 5–5 Effect of temperature on consistency.

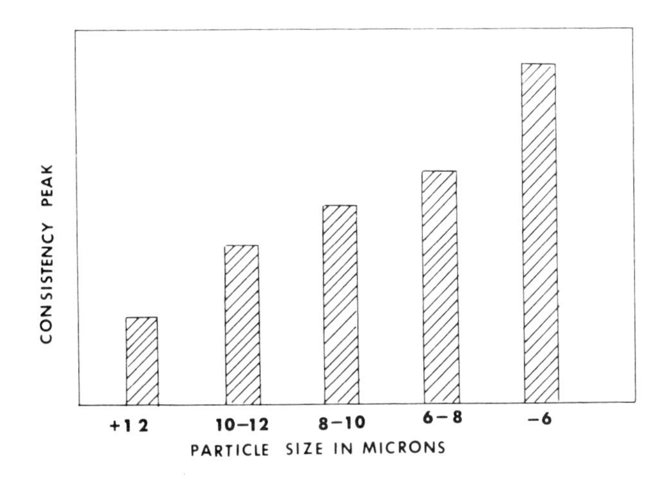

Fig. 5–6 Effect of particle size on maximum consistency of a kaolin.

have been discussed in Chapter 3. Data show that the large monovalent ions reduce the ordered structure of water.[7] This agrees with the observed results which indicate that such systems exhibit low yield points and poor workability as compared with systems containing H_3O^+, Ca^{+2} or Al^{+3}.

The importance of the nature of the adsorbed water has been shown by Grim,[8] as well as Grim and Cuthbert,[9] who base their theory of the bonding action of clay on the nonliquid nature of the initially adsorbed water. Maximum bonding is developed when a restricted, optimum amount of water is added. This amount is believed to be that which develops a definite configuration. Further additions of water result in weakening of the bonds because this additional water results in normal, nonrigid water structure.

On the basis of the previous discussion of water structure and the information presented in Figure 5–4, maximum plasticity is produced in a small-particle-size system where there is a close approach distance between particles, and added ions promote the development of the water structure so that the rigid water structure extends from one particle to another. This explains the action of the lime addition to the wet clay mix to improve plastic properties. It also explains why the addition of a Southern bentonite containing Ca^{+2} ions improves plasticity more than the Western variety that contains Na^+.

Summary

Plastic properties of kaolin-water systems may be improved by:

1. Decreasing the charge on particles by changing the clay to a type that does not have a monovalent cation adsorbed on its surface.
2. Avoiding additions that lower the surface tension of the water phase.
3. Increasing the rigidity of the adsorbed water structure by

avoiding monovalent cations and adding the highly charged ions, such as Ca^{+2}, Mg^{+2}, Al^{+3} or H_3O^+ (acid).

4. Maintaining normal temperatures rather than higher temperatures during the forming process unless previous adjustments in water content have been made.
5. Decreasing particle size or adding fine-grained clay minerals.

REFERENCES

1. West, R., "The Plastic Behavior of Some Clays," Ch. 8, *Clay-Water Systems,* W. G. Lawrence, ed., SUNY College of Ceramics, Alfred, N.Y. 1965.
2. Lawrence, W. G., "Factors Involved in the Plasticity of Kaolin-Water Systems," *J. Am. Ceram. Soc.,* 41, 5: 147–50 (1958).
3. Norton, F. H., "Fundamental Study of Clay: VIII, A New Theory for the Plasticity of Clay-Water Masses," *J. Am. Ceram Soc.* 8: 236–40 (1948).
4. Sherwood, T. K., "Drying of Solids," II, *Ind. Eng. Chem.,* 21, 10: 1976–80 (1929).
5. Westman, A. E. R., "Capillary Suction of Ceramic Materials," *J. Am. Ceram. Soc.,* 12: 589–95 (1929).
6. Kingery, W. D. and Francl, J., "Fundamental Study of Clay: XII, Drying Behavior and Plastic Properties," *J. Am. Ceram. Soc.,* 37, 12: 596–602 (1954).
7. Robinson, R. A. and Stokes, R. H., *Electrolytic Solutions,* Academic Press, N.Y., 1951.
8. Grim, R. E., *Clay Mineralogy,* pp. 132–34, McGraw-Hill Book Co., Inc., N.Y., 1953.
9. Grim, R. E. and Cuthbert, F. L., "The Bonding Action of Clays, I, Clays in Green Molding Sands," Illinois State Geology Survey Report, Investigation 102, 1945.

CHAPTER 6

DRYING

After forming, proper drying is the next important step. It is unfortunate, but not surprising, that many of the properties of clay which make it such a wonderful, plastic material are the same ones responsible for most of the problems associated with drying. The clay particle tends to associate with water and even in the process of drying holds tenaciously to it.

If one starts with a plastic clay containing the proper amount of water to obtain maximum plasticity, the particles will, as previously shown, be separated by water films. If the volume of a piece is measured and recorded as drying proceeds, a drying curve is obtained (Figure 6–1).

Starting at point A, the volume of water removed is equal to the volume shrinkage; hence, the 45° slope of the line from point A to point B. At B the particles are in contact and no further shrinkage occurs. As drying proceeds below this point, the water removed is replaced by air. Because shrinkage is complete at point B and no further movement takes place in the piece, the drying process can be hastened below point B. When movement is occurring in a piece, whether it be due to forming, drying, firing or heat shock, defects are likely to occur. The inserts in Figure 6–1 show the particle arrangements corresponding to the various points on the drying shrinkage curve.

If one plots the rate of drying (Figure 6–2) against the percent water remaining in the piece, one can observe that the rate of drying is constant during the initial stages and then decreases. The constant rate period of drying from point A to B corresponds to removal of water from the surface. As this water is removed the piece shrinks; hence, a continuous water layer is maintained on the surface. At point B, when particles come in contact with each other and shrink-

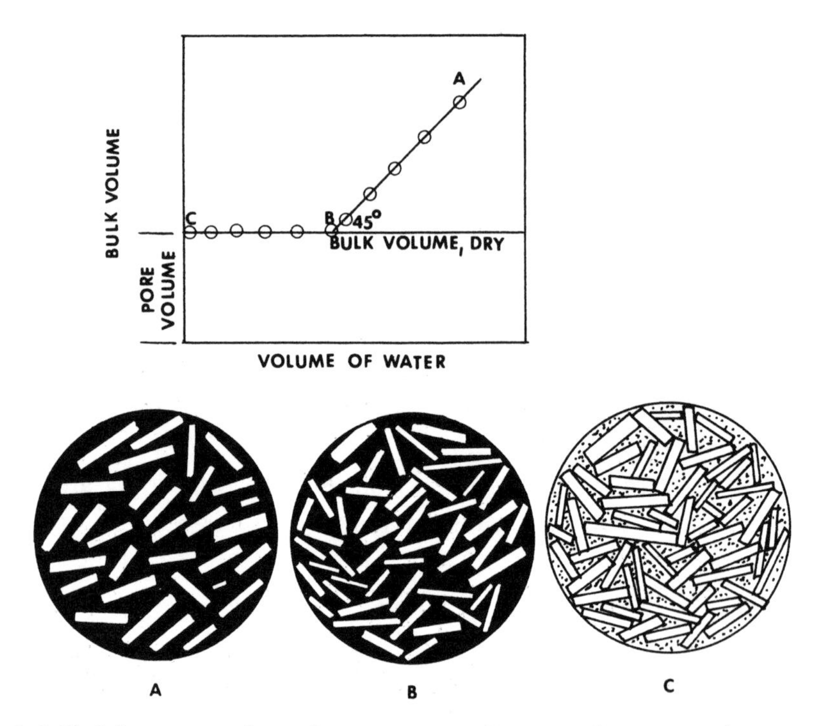

Fig. 6–1 Shrinkage curve for a clay-water paste. Inserts indicate particle arrangements at points A, B and C.

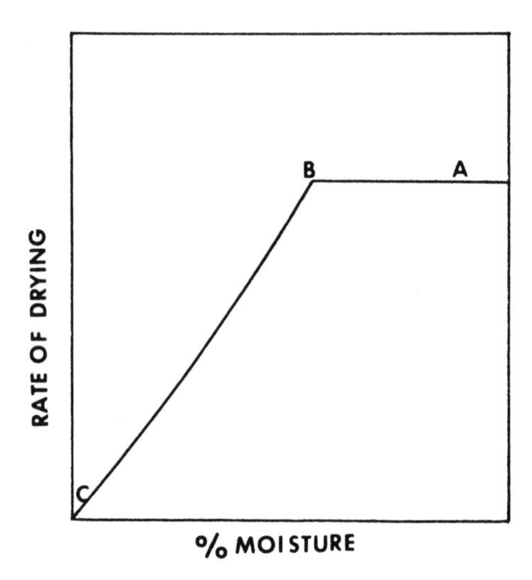

Fig. 6–2 Rate of shrinkage for a clay-water paste. Points A, B and C correspond to those in Figure 6–1.

age ceases, continued drying causes the water surface to retreat further and further beneath the surface of the ware. Therefore, the drying rate decreases.

The term *shrinkage water* is applied to the water that separates particles in the plastic mass and is present in the system from point A to point B.

At point B when particles are in contact and shrinkage ceases, the water that remains is called pore water. The amount of pore water present in any given clay at point B will vary but depends largely on the particle size distribution of the clay.

Some shrinkage curves (Figure 6–3) for various clays are given by Norton.[1] These curves indicate that the finer grained clays shrink most. This would be expected because in such clays there are more

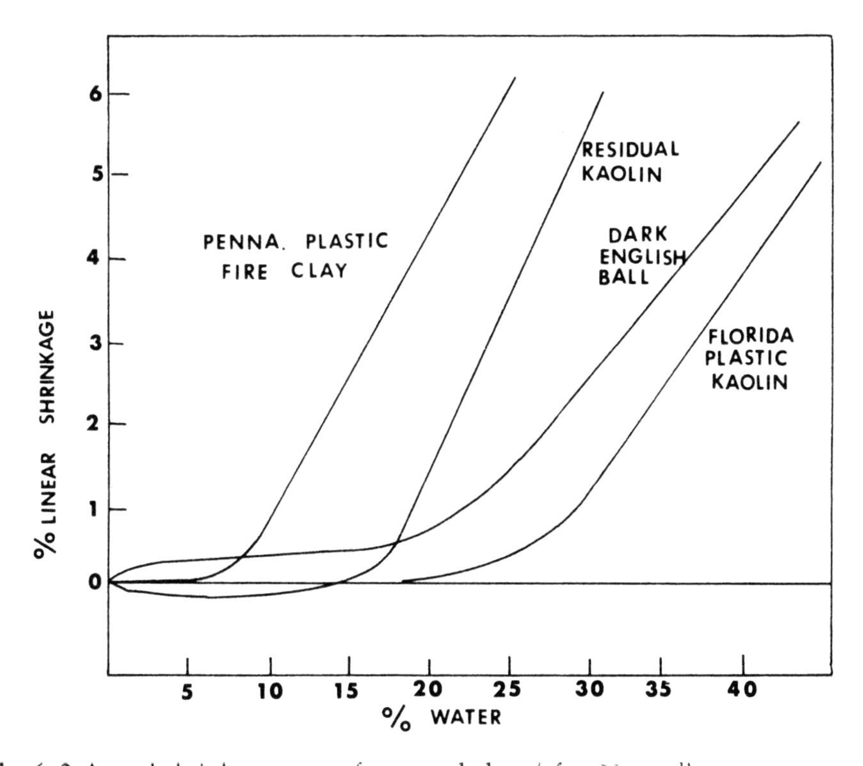

Fig. 6–3 Actual shrinkage curves for several clays (after Norton[1]).

water films per unit length. The curves also show great differences in the water content of different clays at point B where shrinkage is complete. The fire clay has only 10 percent water while the fine-grained kaolin has approximately 25 percent.

Causes of Drying Defects

FORMING IMPERFECTIONS

In any forming operation imperfect flow of the material may occur. This causes discontinuities, such as laps, seams, laminations or other areas, in which perfect continuity of the body is not maintained. These are points of weakness, and, in the drying process, the resulting shrinkage may cause cracks to develop at these points.

CLAY PARTICLE ORIENTATION

The fine size and the platelike particle shape of clay minerals are responsible for drying problems. Because of the platelike shape of the particles, all forming operations tend to result in orientation of these particles with their long dimension perpendicular to the forming pressure.

The shrinkage of the clay is least in the direction parallel to the platelet orientation, most in the direction perpendicular and intermediate in the unoriented condition. The piece shown in Figure 6–4 will tend to warp during drying or if the stresses developed exceed the strength of the body, it will crack.

DIFFERENTIAL WATER

All forming operations, even though uniform water distribution is achieved in the starting mix, result in a condition of differential water content in the ware. In extrusion, the frictional forces between the

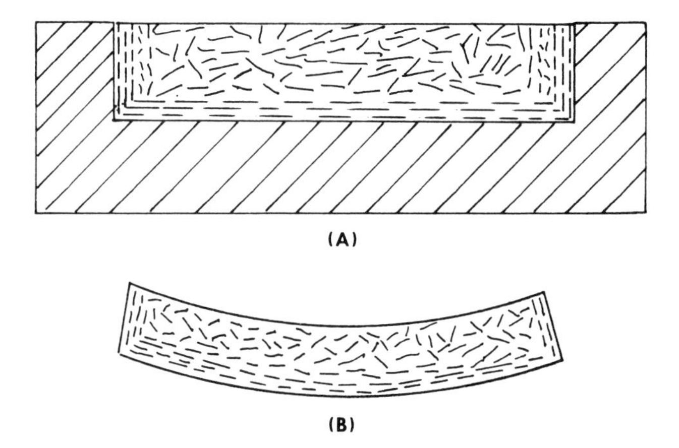

Fig. 6–4 (A) Orientation effect on surface of slip-cast piece and (B) warpage resulting from drying.

material and the die cause a migration of water toward the center of the extruded column. A large slip-cast piece will have a moisture gradient through the wall, the drier side being the one next to the plaster mold. A piece thrown on the wheel will contain more moisture on the thrown surface due to the necessity of using water as the lubricant during the forming process. If water is allowed to stand continuously in the bottom of the pot during the throwing operation instead of being removed occasionally by sponging, it is very likely that a crack will develop in this area during the drying process (Figure 6–5).

The reason for the development of drying cracks caused by differential or uneven water content in the piece can be seen by examination of the drying shrinkage curve shown in Figure 6–6. At the start of drying, one surface of the ware has a high water content, for example, W_2 in Figure 6–6, and the other surface, W_1, has a lower water content. There will be a difference in the amount of shrinkage between the two areas. The portion of the body containing the higher amount of water will shrink more than the drier portion. Because it will be restrained from shrinking by the drier portion, tensile stresses

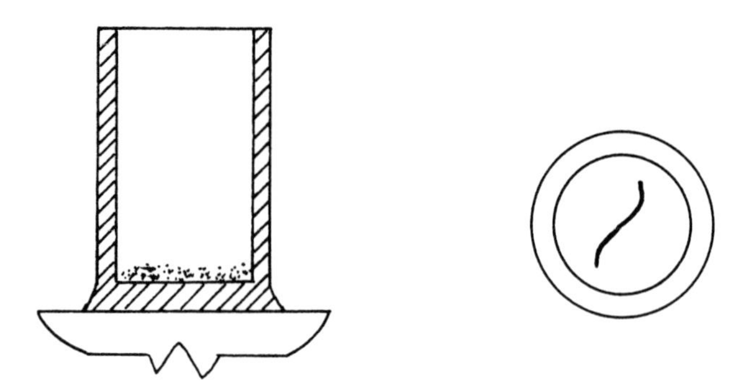

Fig. 6–5 Water collected in bottom of piece during throwing operation (left) resulting in drying crack (right).

will be set up in the portion of the ware that contains the higher water content. Ceramic bodies are weak in tension, and, as a result, a crack will develop in the part that had the higher water content in the beginning.

NONUNIFORM DRYING

Even though a piece may be absolutely uniform in moisture content after forming, it will warp if not evenly dried.

For example, consider a flat piece, such as a tile. If the tile is

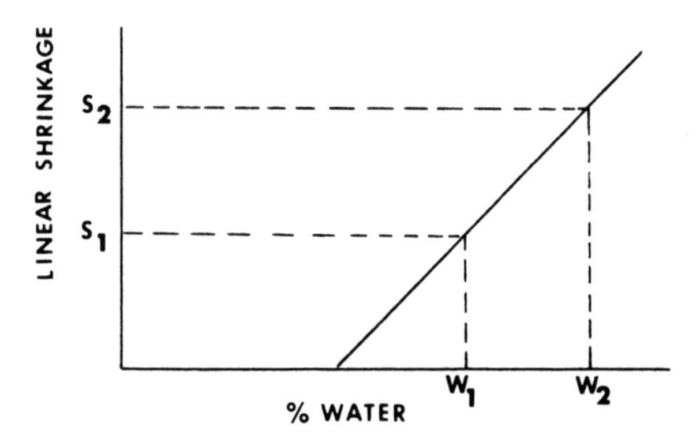

Fig. 6–6 Shrinkage curve for plastic clay illustrating different shrinkage for two different water contents.

placed on a slab or metal plate so that evaporation or drying can take place only from the upper surface, it will warp. The upper surface, as it loses water, will shrink, while the opposite surface will not shrink as fast.

DEVELOPMENT OF MOISTURE GRADIENT IN THE WARE

The drying process itself may result in development of a moisture gradient. During the initial stages of drying, the volume of water removed is equal to the volume shrinkage of the piece. The particles are readjusting themselves as the water is removed, until a particle-particle contact is made throughout the structure. This shrinkage water is being evaporated at the surface of the piece. It is important that this water not be removed from the surface of the ware faster than it can be replenished from the interior. The rate at which it can be replenished depends on factors affecting the migration of water through the small capillaries or interstices in the clay body. Fine-grained clays must be dried very slowly to prevent surface cracking while coarse-grained clays can be dried more rapidly. Bodies containing grog have lower clay content and less shrinkage. Such bodies can be dried faster. An important factor in the migration of water to the surface is the viscosity of water. As water is heated it becomes less viscous. This is the reason for the use of humidity driers that allow the piece to be initially heated in a very humid atmosphere which prevents drying. Then, by reducing the humidity, the rate of surface evaporation can be controlled. Because the piece is heated throughout at the drying temperature before any evaporation from the surface is allowed to take place, there is less chance of developing an undesirable moisture gradient through the piece during the drying process.

SEGREGATION

Segregation, settling or unmixing is of little or no concern in plastic systems because the consistency of the mix is such that the mobility of the ingredients is restricted and segregation cannot take place. However, it becomes important in casting slips, particularly if unusually large or heavy nonplastic materials are used. In such cases, segregation or settling may occur. This may result in the bottom surface of the cast piece containing more nonclay material, such as grog, flint or feldspar, than does the upper portion. As a result the bottom surface would shrink less than the top and warping would occur much like that shown in Figure 6–4.

DRIER SCUM

Drier Scum is not always abhorred by ceramists. Some brick manufacturers use clays or shales high in gypsum to manufacture facing brick with large white splotches called colonial brick which claim a premium price. However, drier scum is normally considered a defect in ceramic ware and efforts are made to eliminate it. Usually small amounts of finely divided barium carbonate are added according to the following equation:

$$CaSO_4 \cdot 2H_2O + BaCO_3 = CaCO_3 + BaSO_4 + H_2O$$

Both the barium carbonate and calcium sulfate are only slightly soluble, but when dissolved in water, they react to form a precipitate that is even less soluble. In dealing with these slightly soluble compounds, the questions arise concerning the amount to add and the method for adding. Certainly intimate mixing is required so that all calcium and sulfate ions have the opportunity of precipitating with the barium and carbonate ions.

The table of compound solubility (Table 4–1) shows that barium carbonate is more soluble in hot water than in cold but that gypsum is less soluble in hot water than in cold. Thus an advantage is gained by adding the proper amount of barium carbonate to hot water and adding the hot water mixture to the clay or shale containing the gypsum. The barium ions are present in as large a concentration as possible so that immediately upon contact with water the ions from dissolved gypsum are precipitated. The barium carbonate has a specific gravity of 4.43 so that the mixture with water must be stirred constantly until it can be added to the clay; otherwise, it settles out very rapidly.

Care must be exercised to add only the proper amount of barium carbonate to the clay to prevent drier scum. Chemical analyses of the clay are time consuming, costly and not always reliable, so a simple laboratory procedure is preferred. Small amounts of dry clay (about 100 g.) should be weighed and then mixed with a series of barium carbonate-water solutions. The first mix should contain no barium carbonate, the second should contain 0.1 percent barium carbonate based upon the weight of the clay, the third should contain 0.2 percent with succeeding batches doubling the amount of the barium carbonate. The clay pastes should be mixed by hand and formed into cubes. Normal drying allows the scum to leach from the interior to cover the surface evenly so that no appraisal can be made of the amount of scum. However, if the cubes are placed on perforated pallets of transite sheetboard, they will dry unevenly with the scum appearing at the holes where drying is fast and no scum appearing in between. The polka dot effect is easily seen and used to evaluate the amount of scum. Drier scum on unfired specimens is clay-colored and not apparent, so specimens must be fired to evaluate the amount of scum.

REFERENCE

1. Norton, F. H., *Elements of Ceramics,* p. 110, Addison-Wesley Publishing Co., Reading, Mass., 1957.

C H A P T E R 7

PARTICLE ORIENTATION EFFECTS

Because the particles of most clay minerals are platelike in shape, they may be expected to become oriented during any forming operation. This is shown for an individual particle in Figure 7–1. The results are the same no matter what means are used to apply the external force used in the forming process—the withdrawal of water from a casting slip by a porous plaster mold, extrusion through a die, plastic pressing, throwing on a wheel, rolling with the hands or slicking a clay bar with a spatula. Some degree of particle orientation will always result and will affect drying and firing behavior.

Before discussing the various individual defects that may be caused by particle orientation, let us review Williamson's work on the effects of orientation on shrinkage.[1–3]

Differential Shrinkage Caused by Particle Orientation

Differential shrinkage can be simply demonstrated by the following experiment. Progressively flatten a wedged block of clay by rolling it out with a rolling pin as in making pie crust. Another simple method is to drop the piece of clay repeatedly on a flat surface turning the clay over each time so that opposite surfaces receive equal impact. Then cut out a test piece and measure the shrinkage in the directions perpendicular and parallel to the plane of rolling or impact. As a result of rolling or impact, the clay particles near the surface become oriented in the manner indicated in Figure 7–2. Measurement of the shrinkage in two directions gives the results shown in Table 7–1.

The ratio of thickness shrinkage to surface shrinkage increases as the piece becomes thinner because a greater percentage of the

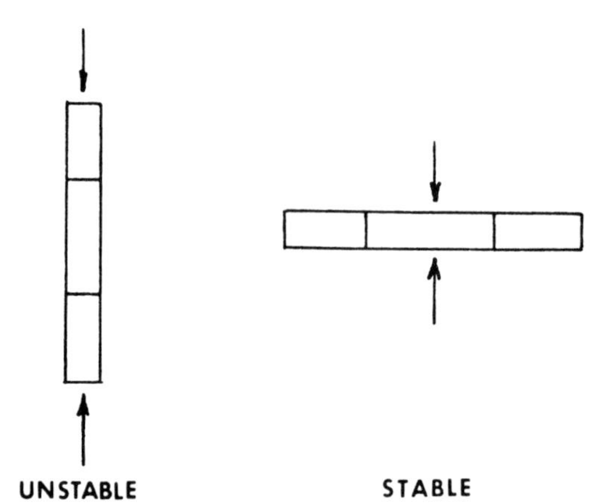

Fig. 7–1 Unstable and stable arrangement of single kaolinite particle with respect to applied force.

particles is oriented. Shrinkage is greater in the direction perpendicular to the particle orientation because in this direction more water films are present per unit:distance.

If the thickness:diameter ratio of a kaolinite particle is 10:1, one could expect a theoretical difference in the shrinkage of 10:1. Complete particle orientation is never achieved and in the experiment shows a ratio of 3.3:1.0. Nevertheless, this difference is significant and great enough to give rise to drying and firing problems.

The difference in drying shrinkage caused by particle orientation may be represented on the familiar drying shrinkage curve shown in Figure 7–3.

Slip Casting

Removal of water from a casting slip by a plaster mold results in orientation of the platelike clay particles with their long dimension parallel to the mold surface. At a sharp corner, as shown in Figure 7–4, there will be a sudden change in particle orientation between the sidewall and the bottom. The sidewall and the bottom will have

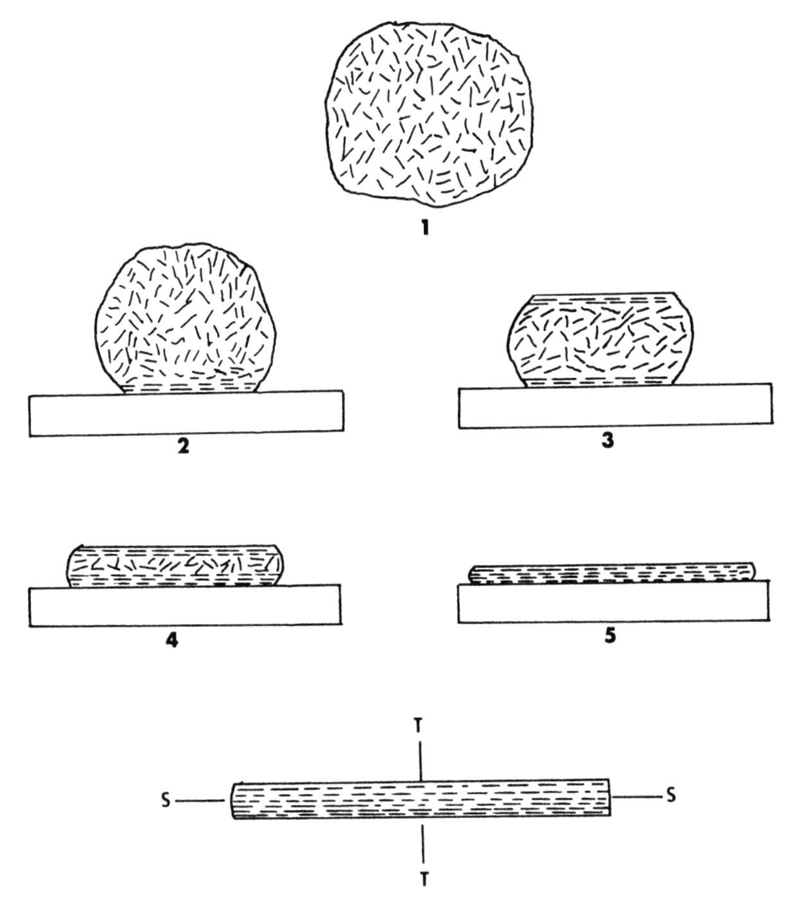

Fig. 7–2 Orientation of particles in a plastic clay mass caused by successive droppings on a flat surface.

their directions of maximum shrinkage at right angles to each other and there will be a strong tendency for the piece to crack during drying at the point indicated.

The remedy is to have more rounded corners so that the change in orientation direction is not as abrupt, and the resulting stresses are distributed over a longer distance.

TABLE 7–1
DIFFERENTIAL SHRINKAGE CAUSED BY ORIENTATION[1]

Thickness (mm.) of Test Piece	Linear Drying Through Thickness (T)	Shrinkage % Parallel to Surface (S)	Ratio T/S
28	12.8	7.6	1.7
20	15.5	6.2	2.5
14	15.1	5.1	3.0
8	16.4	5.0	3.3

Laminations[4–7]

A freshly formed piece of ceramic ware might be described as an accident looking for a place to happen. Usually there are so many weak planes, variations in structure, variations in water content and in drying and firing shrinkage that the ceramic ware only survives the manufacturing operations when drying and firing are conducted under optimum conditions. Such ideal conditions are rare, and the ceramist should pay much attention to the reduction in flaws in the ware during forming. The most common term describing such a flaw is *lamination,* although this term covers a wide range of flaws caused by different forming conditions. A lamination might be described as a weak plane in the ware caused during the forming operation which breaks open during subsequent drying or firing.

One of the most common forming operations that produces laminations is extrusion with an auger. This is not the only method that produces laminations, but the characteristics of the device that produce these flaws are common to other operations. For this reason, laminations will be discussed in terms of the auger.

The auger operation is similar to that of coiling large diameter

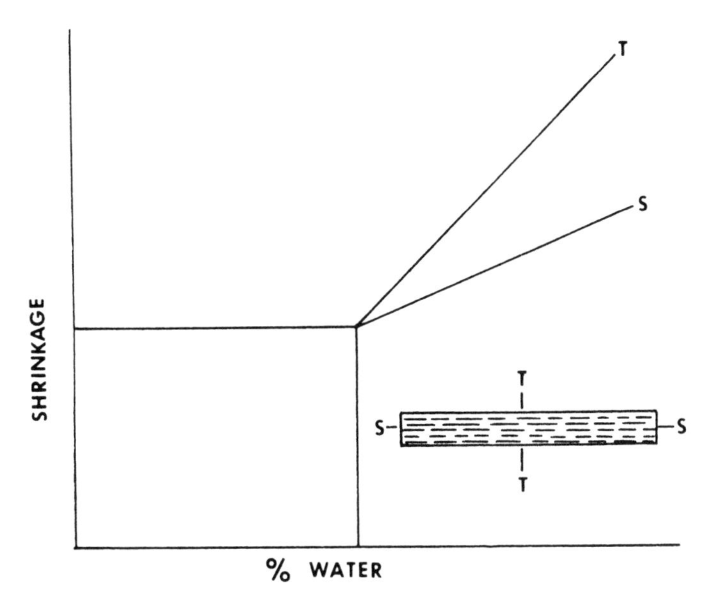

Fig. 7–3 Shrinkage curves in direction parallel to S and perpendicular to T particle orientation.

rope on the ground. A coil of rope is laid in a circle on the ground and then other coils are laid on top. The auger causes a coil of clay to be formed and then pushed through a die with other coils following. The wiping action of the auger surfaces causes preferential orientation of the clay parallel to the auger surface without an interlocking of the clay plates in one coil with those of the subsequent coil. Clay plates at the outer surface of the column are parallel with the surface of the die because of the frictional forces during extrusion. The friction between the clay and the die produces a pressure that causes some of the water to migrate toward the center of the clay column, making it slightly softer with a higher drying and firing shrinkage. Slip planes may also crisscross the column in a manner similar to that of geological faults, also causing a parallelism of clay plates in these weak planes. Obstructions to the passing of clay through some areas of the die may cause a resistance that results in slip planes or

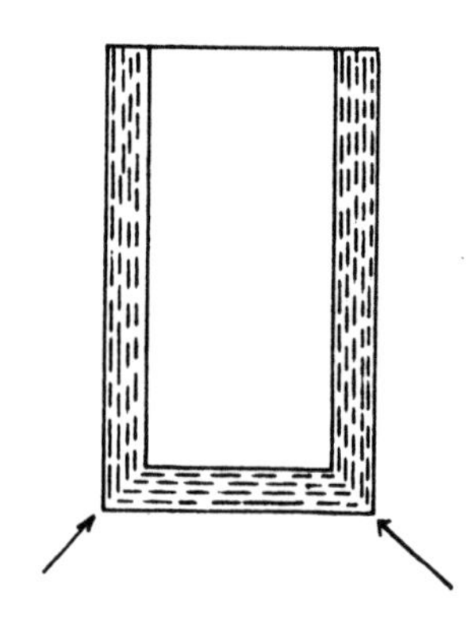

Fig. 7–4 Particle orientation resulting from slip casting and the probable location of cracks caused by sudden change in orientation direction.

surges in the clay column. Objects such as wires, threads, paper or string are serious impediments to the passage of the clay and efforts must be made to preclude their presence.

Any action that has a tendency to separate one layer of clay from another is undesirable because laminations may result. Extrusion is normally performed under vacuum, but vacuum gauges and pumps are frequently inadequate because air leaks in the system often cause the vacuum pump to work over its capacity. The vacuum should be maintained at 26 in. so that a layer of air does not come between the clay layers and produce an impediment for sticking together. The vacuum indicated by the gauge will be different, depending upon the elevation above sea level, because the gauge merely indicates difference in air pressure between the inside of the vacuum chamber and atmospheric pressure. However, this vacuum should be one of the first check points when laminations begin to occur.

Adding water late in the extrusion process or adding water to the filter cake during extrusion produces a water layer between the clay planes that may lead to laminations. Even though the water content of the filter cake is perfectly uniform before extrusion, the pressure caused by frictional forces between the clay and die will produce some water migration toward the center of the clay column which may be damaging.

The rotation of the wing tip of the auger causes a surging action of the clay at any one point of the clay column. That is, the wing tip of the auger, or the last flight, does not push the clay through the die uniformly. As the wing tip rotates, the clay surges forward and then slackens. This surging action may be lessened by using a double wing tip with two flights instead of a single wing tip. The clay column is broken into two separate coils, and the surges are twice as frequent but halved in intensity. A triple wing tip is occasionally used on the auger, and, in this case, the number of clay coils is increased to three and the intensity of the surging is reduced to one-third.

Inspection of the extrusion to assure that the column is moving out evenly is called balancing the die. First, the auger is stopped and the column is cut off evenly with the end of the die. The clutch of the auger is engaged momentarily, so the column extrudes for several inches. Then the column is measured at all points to assure that the clay in any one part of the column is not moving faster than that in another part.

Sometimes a lubricating die is used to lessen the friction between the die and the clay column, but no one lubricant appears to work effectively all of the time. At times steam may work best and at other times some type of oil works satisfactorily.

Laminations are not usually observable in the clay column under normal operating conditions, as they only appear when weak planes break apart during too fast drying or firing. Sometimes a lamination may be identified as an auger lamination when the crack takes an S-shaped form, and vacuum may be intensified or other action taken to reduce the planes in the clay caused by the wiping action of the auger. However, the weak planes in the clay column which may break later into laminations may be determined by two simple tests. Cut out a section of the extruded column and place it directly into a pail of water. The column has a tendency to slake apart at the weak clay planes which may be easily seen. If the weak planes in the clay column are minimal and not easily seen by soaking in water, cut a section from the column and place it in a freezer. The column will break apart at the weak planes where water content is slightly higher.

Orientation Caused by Throwing on a Wheel

Even though the forces used in the hand forming of shapes on a potters' wheel are small, orientation of particles takes place. When pulling up a vertical wall, the fingers are exerting pressure on both sides of the wall. The particles on the surfaces are mobile as a result of the water addition used as lubrication. The particles become oriented with their long dimensions perpendicular to the pressure. Similarly, when the bottom of the pot is shaped, the force is exerted downward or perpendicular to that of the sidewall. This results in the orientation direction of the particles in the bottom being at right angles to the sidewall, and in different shrinkage between the bottom and the sidewall. Thus, a crack is likely to occur at the bottom corner which is the junction between the two orientation directions. The cover of a pot shrinks more across its diameter than the rim of a pot does across its diameter. The rim shrinkage is distributed throughout its circumference which, when translated to diametric shrinkage, is much less. This fact must be taken into account in order to assure a good fit between cover and rim (Figure 7–5).

The Effect of Wedging

Plastic clay is usually prepared from a clay-water slurry from which a portion of the water is then removed either by filter pressing or pouring the slurry into plaster bats where the water is removed by

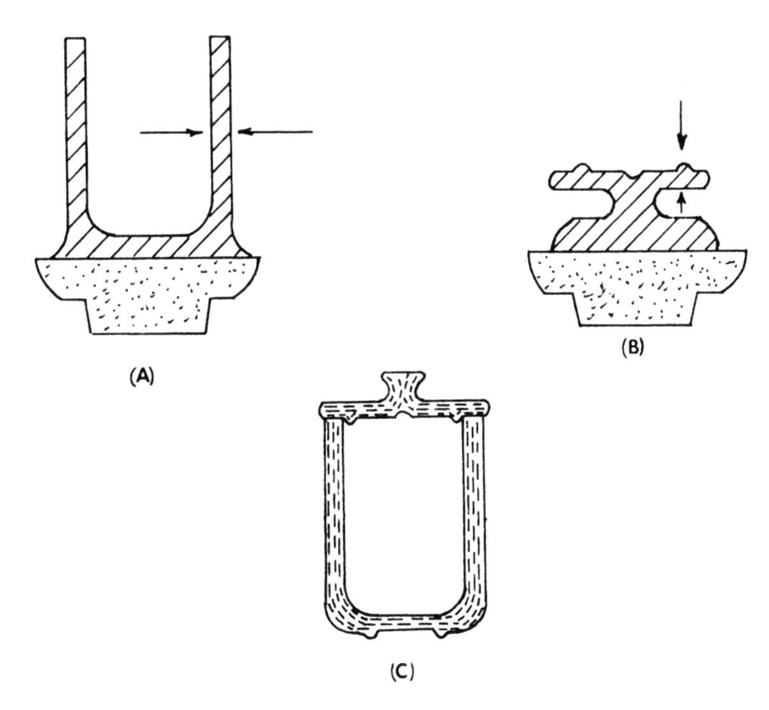

Fig. 7–5 Pressure directions when throwing sidewall (A) and cover (B) with the assembly (C) showing difference in resulting orientation of particles.

absorption. Either method results in a high degree of card pack particle orientation. There is also a nonuniform moisture distribution in the resulting cakes. If these two effects are not minimized, it will be difficult to make a piece free from defects resulting from differential shrinkage.

Wedging serves to randomize the arrangement of particles and to distribute the moisture more uniformly throughout the mass (Figure 7–6). In the wedging process, a small-diameter wire is usually used to cut the plastic mass because this method produces a minimum orientation effect on the cut surfaces. The portions are thrown together and the process repeated over and over. Although complete randomization of orientation is never achieved and localized orientation will always be present, proper wedging minimizes it to the

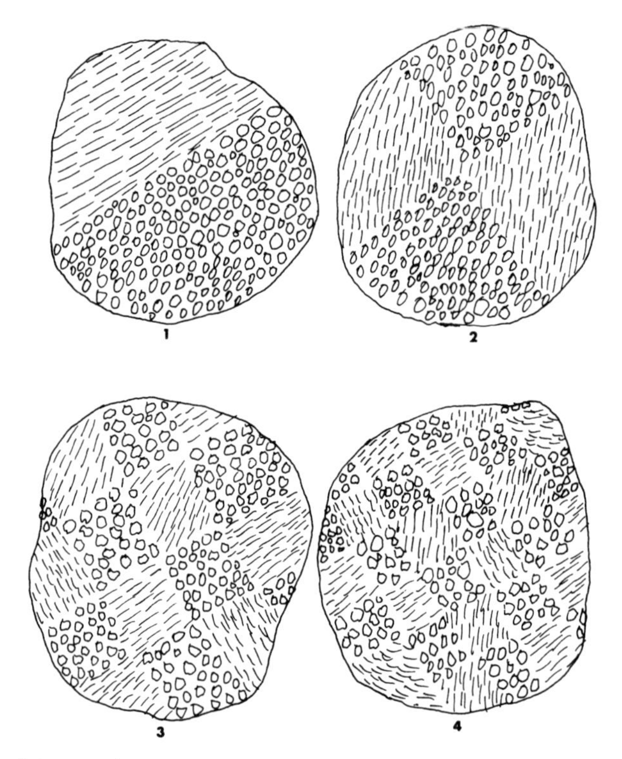

Fig. 7–6 Breakdown of orientation zones as a result of wedging the clay.

point where it is of no consequence. Other functions of wedging are the gradual removal of excess water and air bubbles.

In some instances, wedging is done in such a way as to purposely orient the particles. This is called oriental spiral wedging and is a method used to counteract subsequent work done on the plastic clay in the throwing operation. It also relates to memory of the clay which will be discussed in the next section.

It is well known that if the wheel is rotating counterclockwise, the throwing operation will generate a clockwise spiral twist in the wall of the column. On drying, this spring unwinds slightly and on firing it unwinds even more. Another example of this behavior is the extrusion of a column to be used for electrical porcelain which causes the column to be wound up like a spring. If the holes are drilled in alignment when freshly extruded, they will be as much as 20 to 60° out of line after drying and firing. Spiral wedging is an attempt to compensate for this memory factor, but the amount of memory demonstrated by a clay depends on each individual's treatment of the clay and may become known after many trials have been made. Such elusive factors as the amount of lubrication during the throwing operation, the amount of reduction in wall thickness made on each pass and the speed of the wheel all will have an effect on the amount of twist developed; thus, any solution to the memory problem depends on adjustments made by the individual.

Memory[18–20]

Memory is a perverse characteristic of plastic clays that requires a great deal of patience and understanding on the part of the ceramist. Williamson made an effort to bring some scientific understanding to this problem, but the cause is not completely understood. However, the preferential orientation of the platy clay particles must be the source of this clay behavior.

Apparently the particles assume a fairly stable network orientation in a plastic clay-water mass. The mechanical action on the paste during forming causes the particles to move slightly to an unstable position, and the memory of the stable orientation causes them to reassume this condition during the shrinkage of drying and firing.

Several illustrations of memory will help us to understand how to deal with the problem in practice. The ball of clay used for throwing on the potter's wheel has a stable network of clay particles. The action of throwing causes the particles to be stretched in a helical fashion, with the greatest dislocation of particles being at the top of the ware. After forming, the ware resembles a spring that has been wound, and so during the drying and firing shrinkage, the clay forces cause the ware to unwind so that features at the top of the ware are dislocated from the rest of the ware.

Electrical porcelain bushings are formed on a lathe by turning extruded clay body billets which may or may not have been dried. The extrusion action on the clay billets also causes them to act as wound springs. After turning the billets to the proper contour on a lathe, holes must be drilled at either end, and these holes must be perfectly aligned after drying and firing. Thus, a ceramist marks the ends of a trial billet to determine the misalignment during drying and firing caused by memory, which may be as much as 20 or 30°. Then holes are drilled with a misalignment of the measured amount.

Laboratory experiments have shown that a plastic specimen that has been bent by being picked up at the two extreme ends will have a memory for the straight shape in which it was formed and return partially to that shape during drying and firing shrinkage. The specimen may be bent back to a straight shape before drying. In this case, the memory will be for the bent shape, so it will return partially to this shape during drying and firing. Any straightening must be overcompensated so the specimen will actually become straight during the drying and firing.

Memory is not time-dependent. In one test, a straight specimen was bent and drying started. The specimen began movement toward the straight shape. Drying was halted by enclosing the specimen in a container without further change. Removing the specimen from the container with further drying caused further movement of the specimen toward the straight shape. The specimen was left for long periods in a container without further drying and no straightening was observed; however, as soon as drying was restarted the movement of the specimen continued.

Chipping

The platelike shape of clay particles is responsible for their important properties of workability and dry strength, but it also causes some complex problems. A platelike particle offers greater surface area per mass of particle and also a lubricity that enhances the workability of clay. However, any mechanical action taken on a clay—kneading, extruding, slip casting, jiggering—causes these particles to be preferentially oriented. In slip casting the particles are parallel to the surface of the plaster mold, in jiggering the plates are parallel to the jiggered surface, and in extrusion the particles are oriented parallel to the surface of the extrusion die, as well as to the plane of the auger as it wipes the clay to force it through the die.

Another important feature of mechanical action on a clay-water paste is the migration of the water. Pressure on the clay-water mass causes the water to migrate away from the pressure. Thus, the jiggered surface of a dinner plate or the extruded surface of a clay column has a higher bulk density with a lower water content than the center of the dinner plate or the center of the extruded column.

Ware produced by any ceramic-forming method has a high degree of clay particle orientation along the various planes caused by the mechanical action of forming. Although the original clay-water paste may have had a uniform clay:water ratio, the formed ware has a nonuniform clay:water ratio because of the mechanical forming pressures.

Firing the ware involves many chemical reactions and physical changes, but the general result is that the fired ware consists of mullite crystals and amorphous silica (or glass) as end results of kaolinite reactions surrounded by a glassy phase produced from the melting of feldspars or other fluxing constituents. A filler phase, such as quartz or alumina, may also be present. Most of the mullite present is primary mullite, or mullite that crystallizes from the kaolinite particle and not from the alumina and silica present in the glassy phase (secondary mullite). It does not normally occur in ware manufactured in the United States but may be found in ware manufactured in Europe or Japan, which is fired to a higher temperature (see Chapter 2).

Electron microscope studies of fired kaolinite particles and fired ware show that mullite has a tendency to crystallize as elongated particles in the plane of the kaolinite plates and does not intergrow from one plate to another. Consequently, the kaolinite plates, which are preferentially oriented in the formed ware, cause the fired ware to have a preferentially oriented crystalline structure as well. Layers of interlaced mullite-elongated particles, which are generally parallel to the surface of the ware, are separated by layers of siliceous glass. The interlaced crystalline structure is normally stronger than the glassy phase; also, differences in thermal expansion between the two phases cause stresses parallel to the surface of the ware.

The layered crystalline-glassy structure may be easily separated by an impact blow. The chips that are removed are thin fish-scale-shaped chips and parallel to the surface of the ware where the clay particles have been highly oriented during forming operations. However, to a greater or a lesser degree, there is a general orientation of clay particles throughout the ware, so a fracture of the body follows these orientation planes of mullite crystals. Sometimes a whiteware manufacturer desires to utilize scrap ware by grinding it and adding it to the body formulation. This fired grind is known as pitcher in the sanitary ware industry. Each of the ground particles is a thin fish-scale-shaped particle that will cut the skin like a razor if the hand is thrust into the material. The shape of the ground ware particles is improper for use as "sand" in the Portland cement used to bond the metal cap to electrical porcelain because it does not interlock.

Rotational Rolling[21]

Manipulation of clay-water pastes in different ways often produces unexpected results; for example, forming a ball of clay into an elongated rope-shape. If this operation is done with close scrutiny, the following observations may be made. Take a sphere of a clay-water paste and place it on a smooth surface. Use a flat sheet of plywood, or similar material, to press the clay slightly while rolling. Instead of assuming a football shape, the clay changes to a cylinder with indentations on the ends. Further rolling produces a cylinder with a smaller diameter, and the indentations on the ends proceed farther toward the center of the cylinder. After a very few rolls the cylinder becomes hollow, and it is possible to see through the hole.

This particular result is not restricted to the handling of plastic clay alone but has been used to form seamless copper pipe by rotational rolling. However, the ceramist must remember that with the molding of clay in particular, the unexpected happens frequently. Making coils of plastic clay to form ware may often result in hollow coils.

Differential Firing Shrinkage

Cox and Williamson[3] have shown that pieces that show differential drying shrinkage also show differential firing shrinkage. The firing shrinkage is also greater in the direction perpendicular to the dried clay platelets. This is due to the fact that when the chemically combined water is removed from the kaolinite lattice, as discussed in Chapter 3, the lattice collapses in the direction perpendicular to the platelet surface but retains its structure in the other direction.[22]

One of the effects of firing shrinkage is the development of seams, particularly in slip-cast ware. Regardless of how closely the mold parts fit together, seams will become visible after firing. This is true even though they may be scraped or trimmed off after drying.[23] The effect is illustrated in Figure 7–7.

Three methods are given by Norton[23] to correct this defect.

1. Hammer the seam lightly, causing plastic flow and breaking up of the orientation of the particles in the seam.
2. Bisque fire to a temperature slightly below the maturing temperature and grind off the seam on a wheel. Refire to maturity.
3. Cut a groove from the surface of the seam and refill with plastic clay.

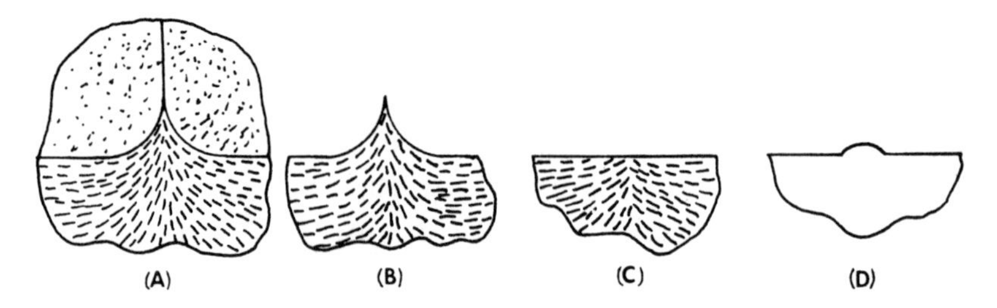

Fig. 7–7 Development of visible seam on fired ware as a result of orientation of particles. (A) and (B) as cast, (C) trimmed, (D) fired.

It is impossible to produce a plastic clay free from particle orientation. Such orientation is greatest in systems that are deflocculated, such as in casting slips, because of greater particle mobility and the charge distribution on the particles in such systems. The smaller the particle size the more susceptible the clay is to orientation. Coarse-grained materials, such as grog, flint and feldspar, will tend to decrease these effects.

Although these effects may not always be troublesome to the potter and although certain traditional practices may help to keep them under control, one should recognize the consequences of particle orientation.

REFERENCES

1. Williamson, W. O., "Causes and Consequences of Clay Particle Orientation in Ceramic Fabrication Processes," Ch. 14, *Clay-Water Systems,* W. G. Lawrence, ed., SUNY College of Ceramics, Alfred, N.Y., 1965.
2. Weymouth, J. H. and Williamson, W. O., *Am. J. Sci.,* 251: 89 (1953).
3. Cox, R. W. and Williamson, W. O., *Trans. Brit. Ceram. Soc.,* 57: 85 (1958).
4. West. R., Fleischer, D., Hecht, N., Hoskyns, W. R., Muccigrosso, A. and Schelker, D. H., "Causes of Chipping of a Stiff-Mud Facing Brick," *J. Am Ceram. Soc.,* 43, 12: 648–54 (1960).

5. Williamson, W. O., "Strength and Microstructures of Dried Clay Mixtures," *Ceramic Processing Before Firing,* G. Onoda and L. Hench, eds., John Wiley & Sons, N.Y., 1978.
6. Moore, F., "The Physics of Extrusion," *Claycraft,* 36, 2: 50–54 (1962).
7. Williamson, W. O., "Microstructures of Plastic or Dried Clay Bodies," pp. 47–57, Proc. Inst. Seminar Clay Miner. Ceram. Processes Prod., Milan, F. Veniale and C. Palmonari, eds., Cooperativa Libraria Universitaria Editrice, Bologna, 1974.
8. Williamson, W. O., "Structure and Behavior of Extruded Clay," *Ceram. Age,* (February, March, April, 1966).
9. Seanor, J. G., "Laminations—What Causes Them in Clay Products," Brick and Clay Record, 142, 4: 80–81 (1963).
10. Robinson, G. C., Kizer, R. H. and Duncan, J. F., "Raw Material Parameters Determining Extrudability," *Am. Ceram. Soc. Bull.,* 47, 9: 822–32 (1968).
11. Robinson, G. C., "Design of Clay Bodies for Controlled Microstructure," I, II, *Am. Ceram. Soc. Bull.,* 47, 5: 477–80 (1968); 47, 6: 548–54 (1968).
12. Reed, A. J., "Auger Design," *Am. Ceram. Soc. Bull.,* 41, 9: 549–53 (1962).
13. Lund, H. H., Bortz, S. A. and Reed, A. J., "Auger Design for Clay Extrusion," *Am. Ceram. Soc. Bull.,* 41, 9: 554–59 (1962).
14. Seanor, J. G. and Schweitzer, W. P. "Basic Theoretical Factors in Extrusion Augers," *Am. Ceram. Soc. Bull.,* 41, 9: 560–63 (1962).
15. Ekedahl, C., "Influence of Water and Pressure On the Forming of Clay Bodies," *Ceram. Age,* 28–33 (May, 1961).
16. Williamson, W. O., "The Fabric, Water-Distribution, Drying Shrinkage and Porosity of Some Shaped Disks of Clay," *Am. J. Sci.,* 245, 10: 645–62 (1947).
17. Weymouth, J. H. and Williamson, W. O., "Some Observations on Microstructure of Fired Earthenware," *Trans. Brit. Ceram. Soc.,* 52, 6: 311–28 (1953).
18. Comeforo, J. E., Fisher, R. B. and Bradley, W. F., "Mullitization of Kaolinite," *J. Am. Ceram. Soc.,* 31, 9: 254–59 (1948).
19. Comer, J. J., Koenig, J. H. and Lyons, S. C., "What are Ceramic Bodies Really Like," I–II, *Ceram. Ind.,* 67, 4: 125–27, 148: 150, 6: 96–99 (1956).

20. Williamson, W. O., "Oriented Aggregation, Differential Drying Shrinkage and Recovery from Deformation of a Kaolinite-Illite Clay," *Trans. Brit. Ceram. Soc.*, 54, 7: 413–42 (1955).
21. ———, Personal communication.
22. Brindley, G. W. and Nakahira, M., "The Kaolinite-Mullite Reaction Series, II, Metakaolin," *J. Am. Ceram. Soc.*, 42: 7 (1959).
23. Norton, F. H., *Fine Ceramics, Technology and Applications*, p. 173, McGraw-Hill Book Co., Inc., N.Y., 1970.

C H A P T E R 8

WHITEWARE BODIES

Whiteware bodies are composed of three essential ingredients: clay, a flux and a filler.

As discussed in Chapter 2, clay provides workability, dry strength, and suspending qualities to keep a slip homogeneous without settling. The clay content is a mixture of ball clay with kaolin which provides a means of adjusting properties. Ball clay is very fine-grained to give exceptional workability and dry strength but has a darker fired color than does kaolin and contains a variety of clay minerals and accessory minerals. Kaolins are normally not as fine-grained as ball clays and consist mainly of kaolinite which fires whiter and produces more mullite which increases the fired strength of the product.

Feldspar and nepheline syenite are the usual fluxing constituents that produce a viscous liquid phase the surface tension of which pulls the solids together into dense ware with firing shrinkage. Auxiliary fluxes, such as whiting, talc or wollastonite, are normally added in small amounts to control the fusion of the body rather than making more drastic modifications in the total feldspar content.

Potter's flint has been the most common refractory material used as a filler in ceramic bodies. In recent years, a substitute has been sought because of the danger of silicosis, which is caused by prolonged exposure to dusts containing quartz in the size range of 10 to 1 μ. Alumina has been used extensively, but its rising costs have made it uneconomical.

The specific composition of a whiteware body depends on its use and the desired physical properties. The following list of sample compositions shows modifications that are made to provide four different types of ware with properties that are considered suitable for the intended use.

	Semivitreous Dinnerware	Hotel China	Electrical Porcelain	Sanitary Ware
Total clay content	53.0	44.0	50.0	48.0
Ball clay	26.0	8.5	32.0	30.0
Kaolin	27.0	35.5	18.0	18.0
Feldspar	13.5	21.0	30.0	32.0
Potter's flint	33.5	35.0	20.0	20.0

Observe that feldspar content is relatively low and flint content high in semivitreous dinnerware and hotel china, whereas the opposite is true with electrical porcelain and sanitary ware. The former are twice-fired with the glaze fire at a lower temperature than the bisque fire. Glazes that fire at the lower temperature require oxides that cause the glaze to have a higher expansion, so the higher flint and lower feldspar content gives a higher expanding body to fit the higher expanding glaze. Electrical porcelain and sanitary ware are once-fired and have glazes that have a lower expansion. Consequently, the flint may be lower to accommodate these lower expanding glazes.

Hotel china uses relatively low clay content with a high ratio of china clay because fired whiteness is extremely important. The ware is usually slightly thicker than semivitreous ware, so dry handling strength is adequate. A great deal of attention is paid to the fired color of ball clays added to semivitreous dinnerware bodies, but good workability is required for fast automatic forming and high dry strength for handling. The fired whiteness of sanitary ware is of little concern because an opaque glaze is used, but the clays must support the nonplastic ingredients in the slip for long periods without settling. The ratio of ball clay to china clay is altered slightly from time to time in order to control casting time.

Reactions on Firing

Discussion of the reactions taking place when clay is heated may be simplified by considering the clay as kaolinite. These reactions were described in detail in Chapter 2 and are summarized as follows:

100–200°C -Removal of free water, completion of drying.
450–600°C -Loss of chemically combined water (hydroxyl).
13.9 percent weight loss.
Loss of crystallinity, formation of metakaolin.
Some shrinkage.
925–980°C -Spinel phase formed with ejection of amorphous silica.
1050–1100°C -Spinel phase crystallizes to mullite with ejection of more amorphous silica. This mullite is primary mullite because it is formed from kaolinite.

The kaolinite loses 13.9 percent of its weight during dehydroxylation and the residue heated to 1100°C (2012°F) gives 64.0 percent crystalline mullite together with 36.0 percent of finely divided and highly reactive amorphous silica. Upon cooling slowly from 1100 to 1000°C (2012 to 1832°F), some of the amorphous silica may change to cristobalite.

Silica

There are three different forms of silica that are of interest to the potter: quartz, cristobalite and silica glass. The thermal expansion of these forms is shown in Figure 8–1.

When heated to a temperature of 573°C (1064°F), the room temperature form of quartz (alpha-quartz or α-quartz) will change to the high temperature form (beta-quartz or β-quartz). This inversion is caused by thermal energy applied to the system which causes the

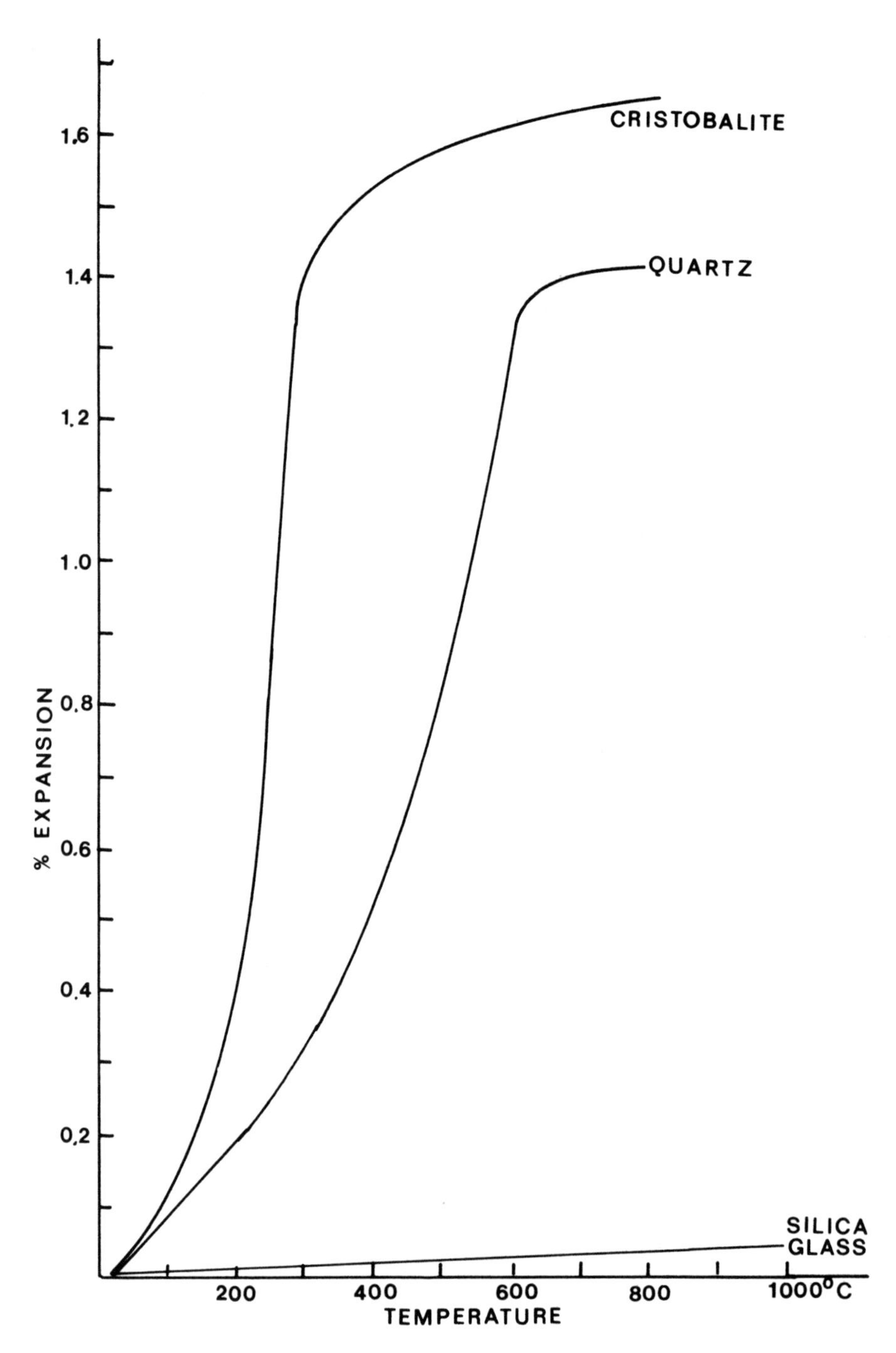

Fig. 8–1 Expansion curves for three forms of silica.

SiO_4^{-4} tetrahedra to rotate with respect to each other. No silicon-oxygen bonds are broken during this inversion; it is a sudden change accompanied by a sudden decrease in size on cooling.

If heating is continued to the 1200°C (2192°F) range, the β-quartz will gradually change to cristobalite, which is another form of silica. This change requires that silicon-oxygen bonds be broken, and a complete rearrangement of ions takes place; it is, therefore, sluggish, time consuming and irreversible. The expansion characteristics of the different forms of silica are important because of their effects on the rate at which a body may be heated or cooled. The ability of a ceramic body to withstand temperature changes is important during the firing process, as well as in subsequent application of the ware in cooking or baking.

At this time, one must differentiate between the silica that is added to the body and the silica that forms as a result of the change from metakaolin to spinel + silica in the 1000°C (1832°F) range. This ejected silica is in the form of extremely small, highly defective crystallites. The form this ejected silica takes seems to depend on the type of clay from which it originated. Some evidence indicates it may be SiO[1], cristobalite or β-quartz. It is fairly certain that it does change to cristobalite during the cooling cycle. If cooled slowly in the 1000°C (1832°F) range a large amount will be formed, while if cooled rapidly in this temperature range, smaller amounts of cristobalite will be formed.

Feldspar

The common types of feldspars used in ceramic bodies are the high potash spars, orthoclase or microcline, $K(AlSi_3)O_8$, and soda spar, albite, $Na(AlSi_3)O_8$. Another mineral closely associated with feldspars and used as a flux is nepheline syenite, $Na_2(Al_2Si_2)O_8$. This is a mixture of 50 percent albite, 25 percent microcline and 25 percent nepheline.[2]

Examination of the phase diagrams for these systems gives an indication of what happens when these materials are heated.[3] When the potash spar, $K(AlSi_3)O_8$, is heated it does not suddenly and completely melt. At 1150°C (2102°F) a high-viscosity glass forms. At this time, orthoclase changes to a different crystal form called leucite, $K(AlSi_2)O_6$, with ejection of SiO_2.

$$\underset{\text{orthoclase}}{K(AlSi_3)O_8} \rightarrow \underset{\text{leucite}}{K(AlSi_2)} + \underset{\text{cristobalite}}{SiO_2}$$

Leucite does not melt completely until a temperature of 1520°C (2768°F) is reached.

Soda spar completely melts at 1118°C, (2044°F), and the resulting glasses have lower viscosities than potash spar glasses at any given temperature.

Another difference between the two types of feldspars involves gas evolution during heating. It has been shown by Hamano[4] that gases are evolved from feldspars when they melt. Apparently the feldspar liquid contains dissolved gases that are released as bubbles while the silica dissolves. Soda spar generates many small bubbles that are trapped in the melt, and potash spar generates a few large bubbles. Because of the high viscosity of feldspathic glass, these bubbles are not easily expelled. This effect is of importance only when translucency is of interest. It accounts for the observation that potash spar melts are more translucent than those of soda spar. In practice, it has been found that there is little advantage of one feldspar over another.[5]

Reactions in Clay-Flint-Feldspar Bodies

After the above brief discussion of the changes that take place when individual components of a typical whiteware body are heated, let us consider a mixture of 40 percent clay, 30 percent quartz and 30 percent feldspar. The clay is usually made up of two or more clays, including kaolin and ball clay. Such a mixture in its green state contains approximately 35 percent pore volume, or air space, between particles.

Brindley and Ougland[6] have made quantitative studies of the reactions taking place in the quartz-kaolin-feldspar systems when heated in the range of 1000 to 1400°C (1832 to 2552°F). Their data for the above composition have been used to construct the curves shown in Figure 8–2. These show the major mineralogical changes taking place when such a mixture is heated, as well as changes in porosity. There are, of course, many other changes and factors involved in the firing of a body which elude graphic illustration. The

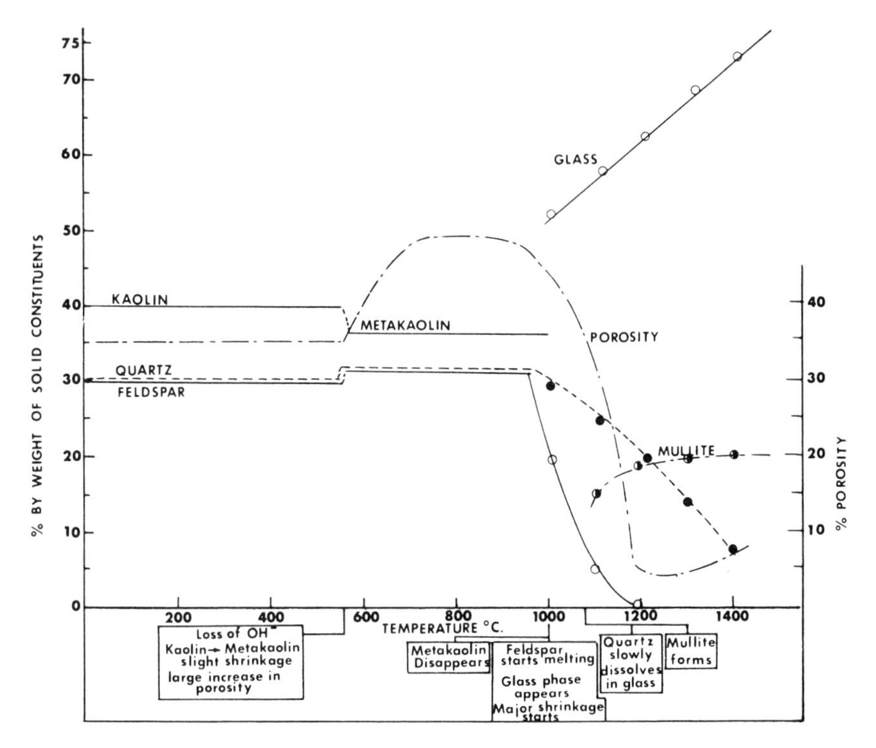

Fig. 8–2 Mineralogical changes in a 40 percent kaolinite-30 percent quartz-30 percent feldspar body when heated. (Data from Brindley and Ougland[6])

work of Brindley and Ougland gives us a good picture of what happens when a body is fired. The temperature ranges in which important changes take place are as follows:

100–200°C (212–392°F)

In this temperature range drying is completed with removal of the last traces of free water adsorbed on particle surfaces. No significant dimensional change takes place.

450–600°C (842–1112°F)

The chemically combined water leaves the clay constituent. The kaolin changes to metakaolin with slight shrinkage but a large increase in porosity. The break shown in the curves for kaolinite, quartz and feldspar reflect the change in percentage of the three constituents resulting from the loss of water from the kaolinite.

573°C (1063°F)

The quartz in the body changes from the low to the high temperature form. Although the expansion curve for pure quartz indicates a sudden, large expansion at this point, it is not too apparent in whiteware bodies because the quartz is fine-grained and makes up only one-third of the body. This expansion takes place at the same time the shrinkage is occurring in the 450 to 600°C (842 to 1112°F) temperature region. These are compensating effects.

300–700°C (572–1292°F)

This is the oxidation period during which impurities, such as organic matter, are oxidized and removed. It is imperative that excess oxygen be present in the kiln atmosphere during this period and that sufficient time be allowed for these reactions to go to completion. Reduction during this period may result in the formation of carbon particles that oxidize very slowly even at high temperatures. If such particles are present after glass formation starts, the trapped gases will cause bloating.

980°C (1796°F)

Metakaolin changes to spinel with ejection of finely divided and highly reactive SiO_2.

1050–1100°C (1922–2012°F)

The spinel formed from the metakaolin starts changing to mullite. There is a very rapid development of mullite needles in the 1100 to 1200°C (2012 to 2192°F) range.

At the same time, the feldspar starts to melt, and a rapid decrease in feldspar content is observed. This is associated with the appearance of the glass phase.

The glass phase starts reacting with and dissolving the finely divided silica ejected from the kaolin during the process of forming mullite. It also begins to dissolve some of the finer quartz grains in the original body composition. Feldspar melts will dissolve large amounts of silica but little or no alumina.

1200°C (2192°F)

At this temperature the feldspar has melted and is no longer detectable by X-ray methods. The glass phase continues to increase. The porosity decreases rapidly and reaches its minimum at this temperature, which is approximately 4 to 5 percent in whiteware bodies. This is due to the formation of closed pores that cannot be eliminated by normal firing methods.

1100–1250°C (2012–2282°F)

Any quartz present changes to cristobalite. West[1] has shown that quartz disappears and cristobalite appears in amounts detectable by

X-ray methods at 1250°C (2282°F) in kaolins and 1200°C (2192°F) for ball clays. The studies of Brindley and Ougland[6] show no cristobalite formed in the three component mixture shown in Figure 8–2. They did, however, show cristobalite for compositions having less than 20 percent feldspar or over 60 percent feldspar. This might be explained by the fact that at the time of the inversion of quartz to cristobalite the newly formed cristobalite is highly reactive. If sufficient glass is present, this cristobalite may be taken into solution, thus preventing accumulation of detectable amounts.

It is known that the presence of fluxes other than feldspar aids in the conversion of quartz to cristobalite. As a result, there is undoubtedly cristobalite present in most bodies fired above 1250°C (2282°F). If considerable amounts are present, precautions should be observed in cooling the kiln. The inversion from the high to low temperature form of cristobalite takes place in the 200°C (392°F) range (Figure 8–1). This is often the range in which the potter becomes anxious to get a glimpse of his work. The kiln door is opened to provide more rapid cooling. If audible "pings" are heard they may be blamed on the cristobalite inversion.

ABOVE 1200°C (2192°F)

Above this temperature very little further development of mullite occurs, although the mullite crystals may grow slightly larger. The porosity increases result from bloating or expansion of the gases present in the closed pores.

From Brindley and Ougland's[6] data the composition of this particular body after firing is shown in Table 8–1.

Thus, a clay-flint-feldspar body, when fired, becomes a mullite-glass-silica composition with the form of the silica, either quartz or cristobalite, being determined by the maturing temperature of the body. The longer the temperature is held above 1200°C (2192°F), the less quartz and more cristobalite will be present in the final composition.

TABLE 8–1
CONSTITUENTS OF A 40% CLAY, 30% SILICA, 30% FELDSPAR BODY

Temperature °C	Glass	Quartz	Mullite
1200	62	21	19
1300	66	16	21

Stresses in Fired Bodies[7–12]

Most of the microflaws in fired ware originate with the quartz particles. A quartz particle dissolving in feldspar liquid causes the release of the dissolved gas in the liquid, so most quartz particles viewed in the microscope show attached bubbles. The quartz particle dissolves to form a very siliceous glass surrounding the particle, and, upon cooling, the stresses formed by differences in expansion (shrinkage) cause microcracks. The silica liquid has a negligible expansion, whereas the quartz particle has an extremely high expansion. Upon cooling, the quartz particle is exposed to high-tension forces because of this differential shrinkage, and most large crystals crack. Quartz crystals smaller than about 30 μ in diameter can withstand the tension and do not crack. A quartz particle approximately 8 μ in diameter will be completely dissolved in feldspathic liquid.

Figure 8–3 shows the stresses formed in feldspathic glass upon cooling below a set point of 700°C (1292°F) caused by differential shrinkage when adjacent to other body constituents. Set point is the temperature at which feldspathic glass becomes so viscous that it acts like a solid and can take stress without deformation. Quartz places silica glass in compression, but siliceous glass places feldspathic glass in tension. As a result, a microcrack usually forms surrounding the quartz particle at a distance of several μ marking the demarcation between highly siliceous glass with a low expansion and feldspathic glass with a high expansion. Lester[10] obtained a patent

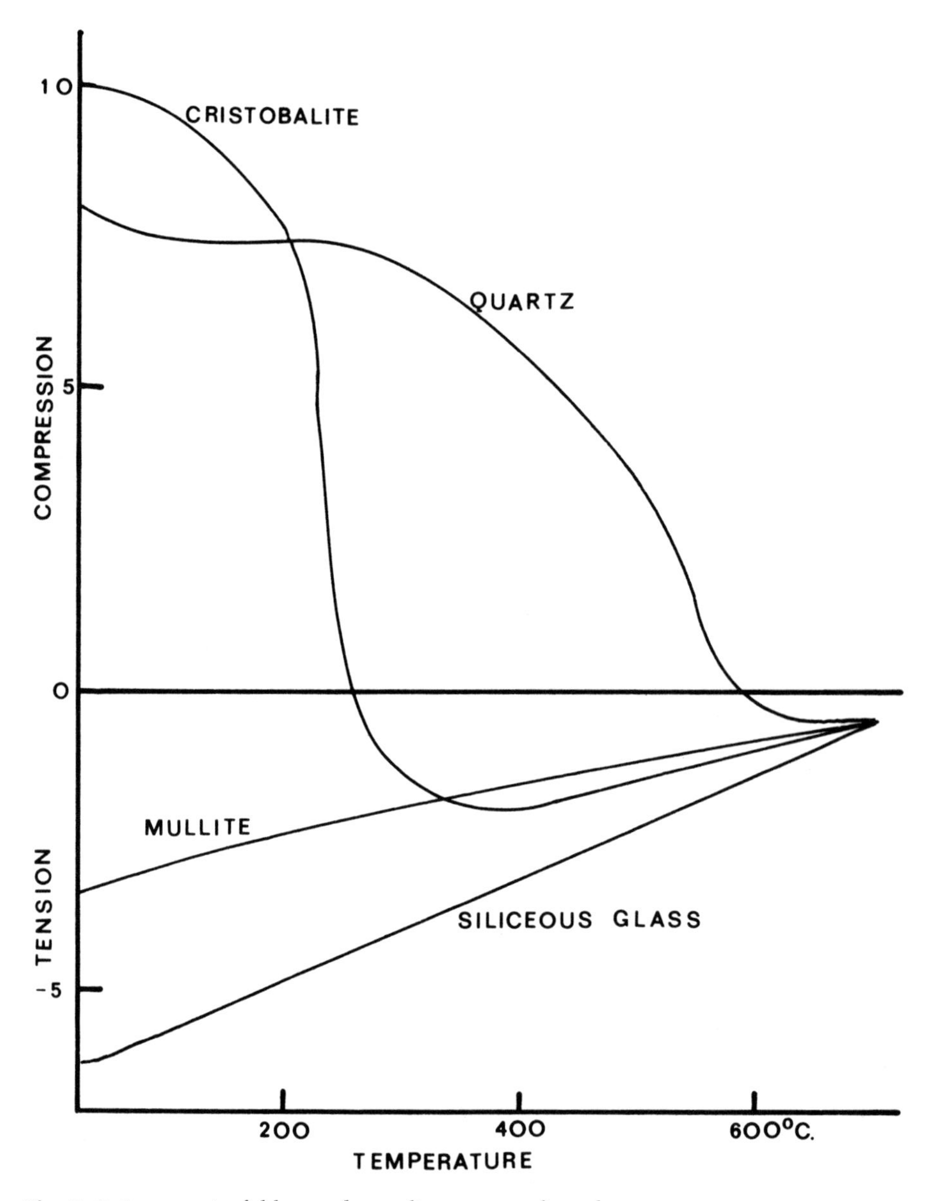

Fig. 8–3 Stresses in feldspar glass adjacent to other phases.

upon a high-strength electrical porcelain body based on minimizing the microcracks. Potter's flint in the size range of 8 to 30 μ produces fewer microcracks because the large particles that crack under stress are not present. The fine particles have been removed because they merely dissolve to form a siliceous glass adding little to the expansion of the body.

The amount of solution of quartz particles depends on four factors:

1. Firing temperature.
2. Firing time.
3. The size of the quartz particles which determines the surface area of the quartz exposed to the feldspathic liquid.
4. The type of feldspathic glass. Glasses from nepheline syenite or those with an auxiliary flux, such as lime, dissolve the quartz more rapidly.

The change in quartz from a crystal to a glass presents a unique problem with firing shrinkage. The quartz filler cannot be interchanged directly by fused alumina or the total firing shrinkage changes. A body containing only alumina as a filler has a firing shrinkage of approximately 15 percent because none of the alumina is dissolved in the feldspathic liquid. Quartz has a specific gravity of 2.65 and silica glass has a specific gravity of 2.20. Thus, when quartz changes to a glass phase, there is an expansion of 20.5 percent. This expansion counteracts the shrinkage caused by the surface tension of the feldspathic glass pulling the particles together, so a body containing potter's flint has a firing shrinkage of approximately 13 percent.

Estimating Expansion of a Fired Body

Once a glaze has been established for continuous use, the modification of the composition causes numerous problems in appearance, as well as chemical and physical properties. Usually the body is more amenable to manipulation for problems of glaze-body fit. This is particularly so when the glaze contains prepared frits. The expansion of a body at a given temperature, such as the set point of the glaze, is the sum of the contributions of each of the constituents based upon the volume percent of the constituent in the body and its expansion at that temperature. There is quite a difference in the specific gravities of the various constituents of a fired body, but this does not detract appreciably from a fairly precise estimate of thermal expansion.

Table 8–2 lists expansions of the more common body constituents at 100°C intervals. If the set point of the glaze is not at a 100° interval the expansions may be interpolated. These expansion values were measured precisely using specially prepared specimens of known composition.

Table 8–3 shows the manner in which the thermal expansion of a whiteware body fired at Cone 10 is estimated. The clay portion might consist of several ball clays and kaolins, but, for the purpose of this calculation, it is appropriate to consider the total clay content as kaolinite with a loss on ignition of 13.9 percent, unless the mineral compositions of the clays have been determined. This is the only constituent of the body that loses a significant weight upon firing and so a correction should be made for the loss on ignition of the clay. The clay residue in the fired body consists of 64 percent mullite with 36 percent siliceous glass. Thus, the contribution of the clay residue toward the total thermal expansion of the fired body is the sums of the expansion for each constituent times the fraction in the fired body. Fused, or calcined, alumina is not attacked by the feldspathic glass in the fired body. Although the feldspars dissolve the silica residue from the clay and also dissolve some of the quartz, the feldspathic glass and the silica glass may be treated as separate glasses for the purpose of this calculation.

The critical portion of the calculations concerns the assessment of the fraction of the potter's flint that is dissolved by the feldspar glass and becomes a highly siliceous glass. A number of different types of whiteware bodies have been evaluated, and, for those using regular potter's flint without further grinding and that are fired to Cone 10, about half of the potter's flint is changed to a siliceous glass.

TABLE 8–2

THERMAL EXPANSION OF CONSTITUENTS IN FIRED BODY

	PERCENT EXPANSION							
	100°C	**200°C**	**300°C**	**400°C**	**500°C**	**600°C**	**700°C**	**800°C**
Silica glass	0.004	0.009	0.015	0.021	0.026	0.031	0.035	0.038
Mullite	0.008	0.050	0.098	0.152	0.206	0.265	0.324	0.387
Alpha alumina	0.050	0.100	0.160	0.250	0.330	0.420	0.520	0.620
Nepheline syenite glass	0.050	0.150	0.240	0.330	0.420	0.520	0.620	0.730
Potash feldspar glass	0.062	0.145	0.228	0.312	0.395	0.478	0.561	0.644
Soda feldspar glass	0.061	0.143	0.225	0.307	0.389	0.471	0.553	0.635
Cristobalite	0.120	0.380	1.390	1.520	1.570	1.610	1.640	1.650
Quartz	0.090	0.190	0.310	0.540	0.830	1.310	1.410	1.410

Note: These thermal expansion values were obtained using standard specimens of each of the constituents heated in a dilatometer.

TABLE 8–3

ESTIMATING EXPANSION OF FIRED BODY

Constituent in Unfired Body	Percent Unfired	Parts Fired	Percent Fired	Constituent in Fired Body	Percent	Expansion of Constituent at 550°C	Expansion Contribution (% × Expansion)
Clay	36.0	30.96	32.6	Mullite	20.86	0.236	0.049
				Silica glass	11.74	0.028	0.003
Alumina	13.0	13.0	13.7	Alpha alumina	13.7	0.375	0.051
Nepheline syenite	14.0	14.0	14.7	Nepheline syenite glass	14.7	0.47	0.070
Soda feldspar	7.0	7.0	7.4	Soda feldspar glass	7.4	0.430	0.032
Potter's flint	30.0	30.0	31.6	Quartz	15.8	1.050	0.166
				Silica glass	15.8	0.028	0.004
Total	100.0	94.96	100.0		100.0		0.375

Set point of glaze = 550°C

Expansion of body at 550°C = 0.375%

Expansion coefficient of body at 550°C $= \frac{0.00375}{550-25} = 7.14 \times 10^{-6}$ in./in./°C

Expansion coefficient equals amount 1 in. (2.5 cm.) of the material expands when heated to the set point temperature divided by the temperature difference between the set point temperature and room temperature.

The whiteware body illustrated in the calculations is a hotel china body fired to Cone 10. The estimated value for expansion coefficient is 7.14×10^{-6} in./in./C° is very close to the measured value and a complete expansion curve plotted from these estimates approximates the measured expansion curve. A glaze having an optimum coefficient for this body would be 10 percent lower or have a value of 6.43×10^{-6} in./in./C°. Thus, it is under slight compression. Often the expansion of the body varies slightly because of faster or slower firing cycles, changes in the amounts of quartz introduced with the clays or feldspars or because of a size change in the grind of the potter's flint. The question arises concerning how much of a composition change should be made in the body to reestablish the coefficient of expansion at the preferred figure. Usually the potter's flint would be increased slightly if the expansion coefficient is too low and an equal amount of the feldspar would be removed from the composition.

Each percentage of potter's flint in the body is responsible for 0.0057 percent expansion in the entire body (0.170/30 = 0.0057%). Each percentage of feldspar, including both the soda feldspar and the nepheline syenite is responsible for 0.0048 percent expansion (0.102/21 = 0.0048%). The replacement of 1 percent quartz for 1 percent clay would be much more effective in changing the expansion as each percentage of clay is responsible for 0.0014 percent expansion in the entire body (0.051/36 = 0.0014%). The amount of clay is usually less easily changed because of workability and dry strength requirements.

Assuming that the expansion of the body should be increased from 0.375 to 0.380 percent, or that the coefficient is to be increased from 7.14 to 7.24×10^{-6} in./in./C°, the amount of potter's flint should be increased 6 percent and the feldspar decreased by 6 percent. However, by decreasing the feldspar content, the effectiveness of the feldspar in dissolving silica has been decreased. In all probability, the first trial in potter's flint increase should be for no more than 4 percent with a later adjustment.

This complication does not arise when increasing the potter's flint at the expense of the clay, and the expansion increase may be accommodated by an increase of potter's flint of only about 1 percent. This is a much less disrupting body composition change than substituting flint for feldspar and the decrease in workability may be accommodated by changing the ball clay:china clay ratio.

Criteria of Ware Quality

The most stringent specifications in the ceramic industry are placed on electrical porcelain because it is expected to have a life of at least 25 years without failure. Other types of ware have varying amounts of open pores, but electrical porcelain is considered to be soft ware if fragments of the ware show any dye penetration into the surface after exposure to fuchsin dye at a pressure of 2000 psi for 16 hours. Each piece of ware is subjected to a high voltage electrical test and a mechanical strength test at half the rated load.

Hotel china ware is tested continuously for impact and chipping strength but is not tested as rigorously for open porosity. Usually samples of ware are tested for water absorption after boiling for five hours, and this is maintained at less than 0.5 percent. However, the centers of some thick ware, such as platters, may show slight ink staining. Fired whiteness is extremely important, as is shape and size perfection. Great attention is paid to color control and application of designs.

Dry handling strength is extremely important with sanitary ware, as is size, shape and glaze perfection. Also, much attention is paid to slip casting control.

Aesthetic properties, such as fired whiteness, color and design application, shape of ware and the economy of production are the major criteria of quality for the semivitreous dinnerware industry. The fired porosity of the ware is maintained at about 6 percent for economy of production with ease in producing perfectly shaped ware. Emphasis is placed on producing new and pleasing designs.

REFERENCES

1. West, R. R., "High Temperature Reactions in Domestic Ceramic Clays," *Bull. Am. Ceram. Soc.*, 37: 6 (1958).
2. Norton, F. H., *Fine Ceramics, Technology and Applications*, p. 81, McGraw-Hill Book Co., Inc., N.Y., 1970.
3. Levin, E. M., McMurdie, H. F. and Hall, F. P., *Phase Diagrams for Ceramists*, p. 154, Am. Ceram. Soc., 1956.
4. Hamano, K., "Studies on the Microstructure of Porcelain Bodies, VII, Origin and Development of Pores in Feldspathic Glasses," *Yogyo Kyokai Shi*, 65, 735: 44 (1957).
5. Geller, R. F. and Creamer, A. S., "Investigation of Feldspar and Its Effect in Pottery Bodies," *J. Am. Ceram. Soc.*, 14: 20 (1931).
6. Brindley, G. W. and Ougland, R. M., "Quantitative Studies of High Temperature Reactions of Quartz-Kaolin-Feldspar Mixtures," *Trans. Brit. Ceram. Soc.*, 61: 599 (1962).
7. Selsing, J., Ceramic Products, U.S. Pat. 2,898,217, August 4, 1959.
8. Chu, G. P. K., *Ceramic Microstructures: Their Analysis, Significance and Production*, p. 844, R. M. Fulrath and J. A. Pask, eds., John Wiley, Sons, N.Y., 1968.
9. Williamson, W. O., "Bubbles in Ceramic Systems," *Ceramurgia International*, 2, no. 1 (1976).
10. Lester, R., Ceramic Product and Method of Making Same, U. S. Patent No. 3,097,101, 1963.
11. Brindley, G. W. and Nakahira, M., "The Kaolinite-Mullite Reaction Series: I, II, III, *J. Am. Ceram. Soc.*, 42, 7: 311–24 (1959).
12. Rowland, D. H., Glazed Insulator and the Like, U. S. Pat. 2,157,100, May 9, 1939.

CHAPTER 9

FIRING

A kiln is not just a box in which ware is heated, but rather a high temperature reaction chamber in which temperature, time and atmosphere all play important roles in the development of the final product. Lack of attention to any one of these variables can result in disappointment. The environment to which a ceramic material is exposed during firing is the kiln atmosphere. The nature and temperature of the kiln gases affect the physical and chemical changes that take place in and on the surface of the ware. Therefore, it is necessary to understand some of the elements of combustion.

Combustion[1]

Much attention is being paid to fuel conservation[2], but this is not economical should the ware be fired improperly. Recent burner developments[3] allow the ware to be fired rapidly while avoiding most of the problems in fired ware, including bloating, black coring and preheat cracks. A thorough understanding of combustion is necessary to take advantage of the economies of firing. Because natural gas may be the most common fuel for firing ceramics, the discussion will concern this fuel, although the same technology applies to other fuels that are used.

Natural gas consists primarily of methane, CH_4, and undergoes the following reaction when mixed with air and burned:

$$\underset{}{CH_4} + 2\underbrace{(0_2 + 4N_2)}_{\text{Air}} = CO_2 + 2H_2O + 8N_2$$

Relative Volumes	1	10	1	2	8

Air consists of approximately 20 percent oxygen with 80 percent nitrogen, so this ratio is normally used to simplify calculations. Thus, when one part by volume of natural gas (mostly methane) is mixed with ten parts of air and ignited, the products consist of one part carbon dioxide with two parts of water vapor and eight parts of nitrogen. This situation is called perfect combustion. Burner equipment is used to proportion the volumes of the gases, and, when the amount of fuel in the mixture is less than that required, it is a lean mix or an oxidizing atmosphere. Mixtures with excess fuel are rich mixes or reducing atmospheres.

Air mixed with methane and admitted at the burner is called primary air, and other air which may be admitted to the kiln through openings in the kiln walls is called secondary air. Kilns are presently being built with little access to secondary air, and the pressure inside the kiln is maintained above the surroundings so that close control may be maintained over all air entering the kiln.

Methane has a heating value of approximately 1000 Btu/ft.^{3} and requires a minimum temperature of 632°C (1170°F) to ignite. A mixture of methane with air containing less than 5 percent volume of methane or more than 15 percent volume of methane will not burn. Mixtures of methane with air in the proper stoichiometric ratios for perfect combustion have a flame temperature of 1918°C (3484°F). This is the highest theoretical temperature that can be attained with this fuel, but higher kiln temperatures may be attained by replacing the air with oxygen alone or by preheating the air before mixing with the methane.

The flame in a mixture of methane with air travels at a speed of 1.5 ft./sec. The function of the nozzle on a burner is to make sure that the methane-air mixture is traveling at a velocity of slightly more than 1.5 ft./sec. so that the flame does not flash back into the burner piping. Too high velocities of the methane-air mixture may cause the flame to blow-off the burner, provided it is not continuously ignited with a pilot light.

Monitoring of the ratio of methane:air is usually accomplished

TABLE 9–1
% CO_2:EXCESS AIR RELATIONSHIP IN COMBUSTION GAS

Ft.³ Air Per Unit Fuel	% CO_2	% Excess Air	% Theoretical Air
10.0	12.0	0	100.0
11.0	10.7	10.0	110.0
12.0	9.8	20.0	120.0
14.0	8.3	40.0	140.0
16.0	7.2	60.0	160.0
18.0	6.3	80.0	180.0
20.0	5.7	100.0	200.0
25.0	4.5	150.0	250.0
30.0	3.7	200.0	300.0

with meters that record the amount of carbon dioxide in the combustion gas. Table 9–1 illustrates this relationship.

The three modes of heat transfer are all involved in the kiln firing process. The outside of the ware becomes hot first and the heat is transferred to the interior by conduction. Below the temperature at which ware glows red, the major means of heat transfer to the ware is by convection, whereby the hot combustion gases circulate through the ware setting allowing the heat in the gases to be transferred to the ware. To allow efficient transfer of heat from gases to ware, movement of the gas must be turbulent so that surfaces of the ware are swept by fresh hot gas. Above the temperature where the ware glows red (about 600°C or 1112°F), the transfer of heat to the ware is by radiation, and this radiant heat travels in straight lines much as light does. The radiant energy may be interrupted by the setting of the ware; therefore, the ware should be set so that one piece of ware does not shadow another piece and prevent it from receiving the radiant heat energy.

Different combustion conditions are maintained in different temperature ranges in firing. In the 550 to 650°C (1022 to 1202°F) range, the ware is heated slowly with large amounts of excess air to oxidize properly the interior of the ware to prevent bloating, black coring and preheat cracking. Special excess air burners (XS air burners) have been developed to heat the ware and to provide the following special conditions required by the ware in this range:

1. Turbulent atmosphere to heat the ware efficiently by convection.
2. Rapidly moving gases that remove water vapor from the surface of the ware to allow the surface to be heated more efficiently.
3. Large amounts of oxygen that may penetrate the pores of the ware and allow proper oxidation of carbon as well as iron and titanium oxides that are susceptible to changing to dark-colored suboxides under reducing conditions.
4. Large amounts of excess air to dilute sulfur gases emanating from the ware and to prevent their condensation on the ware that is cooler.

Large amounts of excess air up to 200 or 300 percent are used in this heating zone of the kiln, and these especially designed[2] burners provide heat penetration of from 12 to 14 ft. from the burner. The highest temperature in the combustion gases is about 12 in. from the burner, and at a farther distance the temperature uniformity of the gases is enhanced.

Above a temperature of 600°C (1112°F) where heat transfer is mainly by radiation, different burners may be utilized to provide the conditions required by the ware in this temperature zone. The excess air is usually throttled back in this zone so that a more quiescent condition is maintained, and the kiln dirt is kept at a minimum to prevent it from clinging to the glaze which becomes fluid in this temperature range. Instead of the penetration burners used in the low temperature zone, a flat flame burner may be used to provide maximum radiation. This type of burner is so designed that the flame

swirls close to the kiln wall and heats it to radiate to the load. The hot gases have no velocity in the direction of the ware being heated. The wide wall expanse that radiates heat minimizes the shadowing effect of the ware in the setting.

Atmosphere Controlled Reactions[4]

$C + O_2 \rightarrow CO_2$ This is the reaction involved in the burning off of the organic matter present in raw materials. The reaction starts at 350°C (662°F) but does not proceed rapidly until a temperature of 600°C (1112°F) is reached.

$4FeO + O_2 \rightarrow 2Fe_2O_3$ Excess oxygen maintains iron oxide in the ferric oxide state. This is the red shade of iron oxide but color varies from orange to purple, depending on the temperature to which it is fired (lighter colors are produced at the lower temperatures). This form gives cream colors to whiteware, buff color to firebrick and red colors to building brick. Ferric oxide is quite refractory and chemically inert, and therefore, often remains as a component in the final product.

$Fe_2O_3 + CO \rightarrow 2FeO + CO_2$ Reduction of the red ferric oxide to the black ferrous oxide will result if carbon monoxide is present in the kiln atmosphere. Carbon monoxide results when there is insufficient oxygen present for the complete combustion of the fuel. Ferrous oxide acts as a powerful flux in a body. It will form glass in contact with siliceous materials at temperatures below 1100°C (2012°F). These glasses are usually green to black and have very low viscosities. Ferrous oxide is responsible for the blue-white color of a whiteware body fired in the presence of carbon monoxide. It also plays an important role in black coring and bloating to be discussed later in this chapter.

Color control The color of many bodies and glazes in which transition elements are used as coloring ions depends on precise atmosphere control. Ions, such as copper, nickel, iron, vanadium, and chromium will change the color of silicate melts when their valence is changed. Thus Cu^{+} is red but Cu^{+2} is green.

Water vapor is a product of the combustion of hydrocarbon fuels and is, therefore, present in the kiln atmosphere. Water vapor apparently acts as a catalyzer for the reactions taking place in the body and may reduce the viscosity of the glass phase formed.[5] Very noticeable differences may be observed between identical products fired in an electric and in a gas kiln. Assuming identical heating schedules, the same degree of maturity is not reached in the electric kiln.

TIME

Reactions taking place in ceramic bodies require time. The softening of pyrometric cones depends on the rate at which they are heated; hence, these rates are specified. In a study of the effect of time on the maturing of a whiteware body, Norton[6] shows that a tenfold increase in time permits the lowering of the maturing temperature by 23°C (41°F).

Fast firing of ceramic bodies is of great interest because of the obvious economic advantages. Present day kilns place no limit on the rate at which a piece may be heated or cooled. The ware itself dictates the firing schedule.

As noted previously, whenever movement takes place in a ceramic body there is the possibility of warping or cracking. Sufficient time must be allowed for the oxidation and decomposition reactions to take place. A third limitation is the time required for the desired crystal phases to develop in the final product.

These time requirements can be minimized by proper body design. Following are some steps that will permit faster firing:

1. Minimize ball clay content consistent with the plastic requirements of the forming method used. This will decrease the time required for oxidation of the organic matter. Because the ball clay is also the finest grained material in the body, using less of it will also increase the porosity of the body and decrease its firing shrinkage.

2. Add inert ingredients, such as grog, to the body. This "opens" the body permitting easier escape of gases during oxidation, dehydroxylation of the clay and reduction in firing shrinkage. A tile composed of inert wollastonite grain bonded with a minimum amount of clay can be fired in a few minutes.
3. Reduce the cross section of the piece. A thin walled piece can be fired much faster than one with a thick wall. The thin wall more readily allows the escape of gases generated within the piece, permits the diffusion of oxygen to the interior and develops a lower temperature gradient. The thin piece will stand the greater thermal shock involved in fast firing and cooling.

Fast firing produces a more fluid glass phase because the solution of silica is minimized. It also results in the formation of more closed pores because time is not allowed for the escape of the entrapped gases. Mullite development is not as pronounced in fast-fired bodies. Several studies have shown that fast-fired bodies have suitable properties as long as the body can withstand the physical stresses of the fast-firing operation.

Firing Defects

Firing defects have been summarized by Jarrett.[7] There is a great variety of such defects, but each has its own discernible characteristics that make intelligent analysis and correction possible.

UNDERFIRING

Among these defects the most common are oversized dimensions, high porosity, off color and lack of ring, indicating underdevelopment of the glassy phase. In glazes underfiring may result in eggshell surfaces, lack of even coverage or, in extreme cases, incomplete melting.

The obvious corrective measure is to raise the temperature or increase the firing time. If such defects occur in only part of the kiln, an uneven heat distribution should be suspected. The kiln should be surveyed to determine what the temperature distribution is. This can be done by the judicious placement of cone plaques.

OVERFIRING

Overfiring results in undersize dimensions, too low absorptions, slumped or deformed ware and undesirable colors. In glazes, overfiring may result in off colors, loss of gloss, running or complete misfit between glaze and body. Again the obvious cure is to lower the temperature, shorten the schedule and ensure a uniform temperature distribution.

DEFECTS CAUSED BY KILN ATMOSPHERE

Probably the most common defects caused by kiln atmosphere are black coring and bloating, which are closely related. As has been noted, it is important that excess oxygen be present in the kiln atmosphere in the 350 to 700°C (662 to 1292°F) range. This is necessary for the oxidation and removal of any carbonaceous materials present in the body. Although oxidation starts at 350°C (662°F), it does not proceed rapidly until 600°C (1112°F) is reached. If oxidation is not complete by the time glass formation begins, the gases generated will be trapped in the glass and bloating will result.

If oxidizing conditions are not good and if the body contains iron oxide, another factor enters the picture and compounds the problem. The carbon present may react with the insufficient oxygen to form carbon monoxide. This gas will reduce the ferric oxide to ferrous oxide, a powerful flux that reacts with its siliceous surroundings to form glasses below 1100°C (2012°F). This results in the formation of a black core in the interior of the body. Once this happens, it becomes very difficult to burn out the remainder of the organic matter. It is confined in the black glass formed, which greatly hinders the diffusion of oxygen to the interior of the piece. Any oxidation that does take place will generate gas that is confined and may result in bloating.

These effects are most pronounced in dense, tight bodies having a large cross section. Excess oxygen and slow firing in the 350 to 700°C (662 to 1292°F) range eliminate this problem. If body reduction is desired, it should be done after this temperature range is past.

CRACKING CAUSED BY FAST REMOVAL OF WATER

The most common defect of this type is the hairline crack, which occurs when the free water is removed too fast. In extreme cases, the piece may explode. If the steam generated from the water cannot escape through the piece, the pressure generated will disrupt the body. Water in this amount should not be present if proper drying preceded firing. All free water may not be removed by drying, or dried ware may pick up moisture during storage under humid conditions. A quick check for the presence of free water is to break a piece of ware and quickly hold the broken surface next to a mirror. If water is present, the mirror will fog up.

The obvious remedy is to improve drying or minimize storage in humid conditions. If this is not feasible, a slowdown in the initial stages of firing is necessary. Open bodies are much less susceptible than dense ones. A plastic body or one containing grog would be much less susceptible to this defect than a slip-cast body in which the particle packing is much tighter.

SHRINKAGE CRACKS OR WARPING

A typical shrinkage curve for a body is shown in Figure 9–1. It is apparent that shrinkage increases up to the point of vitrification, V. Below this point, shrinkage increases rapidly with increased temperature. This means that in the temperature range where shrinkage is taking place rapidly, care should be exercised in firing to assure a uniform temperature distribution in the kiln. If the temperature dis-

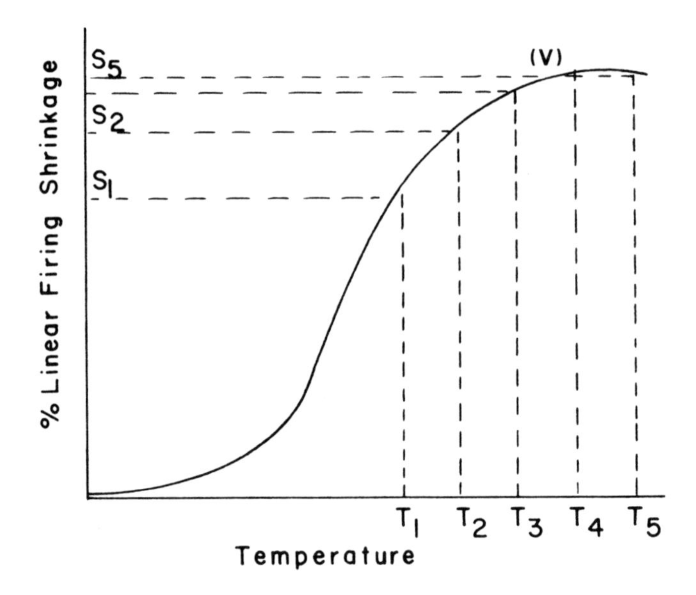

Fig. 9–1 A typical shrinkage curve.

tribution in the kiln is such that the top of a piece is at T_2 while the bottom is still at T_1, the top will have shrunk, S_2-S_1, more than the bottom (Figure 9–2). This may result in the opening of a crack in the top area. Once the temperature has reached the point where the rapid shrinkage is complete, temperature differences will have little effect, T_3, T_4, T_5.

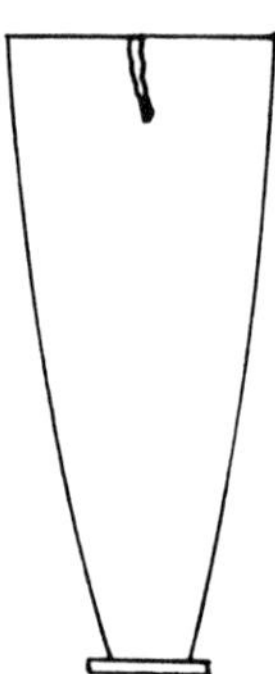

Fig. 9–2 Crack resulting from uneven temperature distribution from top to bottom. The top, being hotter, shrank earlier than the bottom and restraint at the bottom developed a crack to relieve the stress.

Warping is the result of a milder case of differential firing shrinkage. The stresses developed are sufficient to deform the body but not crack it.

But other factors may account for the warping and cracking; improper temperature distribution cannot always be blamed for these defects. Density variations resulting from the forming process, segregation, preferred orientation, flow under the pull of gravity (slumping) and friction between the base of the piece being fired and the kiln slab will all contribute to warpage and cracking during the firing operation. An abrupt dimensional change in the cross section of a piece may also lead to crack formation at the junction because the thicker cross section will shrink at a rate different from that of the thinner section (Figure 9–3).

BODY CRACKS

Most ware contains many incipient flaws relating to particle orientation, moisture gradient, memory or manufacturing-related problems that under any but the most favorable drying and firing conditions will cause breakage. This type of loss should be isolated and corrective measures taken at the stage of manufacture under question, rather than wasting time on modifying the drying and firing schedules. Figure 9–4 shows a typical example, spiral cracks that appear after the firing operation. These are caused by the twist during throwing, resulting in planes of weakness that open up during firing because of the memory of the clay. This may be minimized by proper wedging, improving the plasticity of the clay to allow more defor-

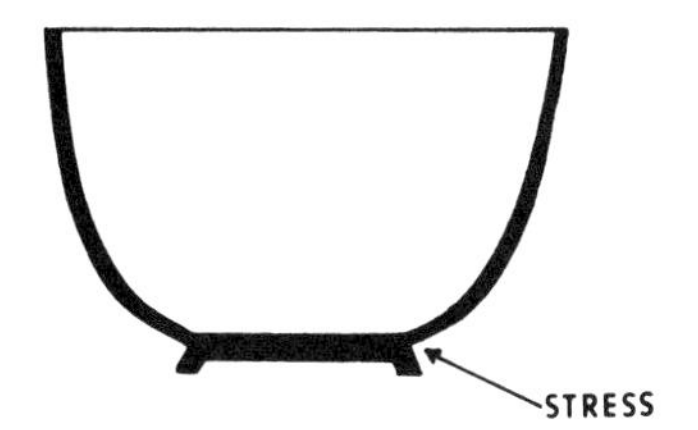

Fig. 9–3 Point of stress concentration as a result of uneven cross sections. A crack is likely to develop at the junction of the two different cross sections during drying or firing.

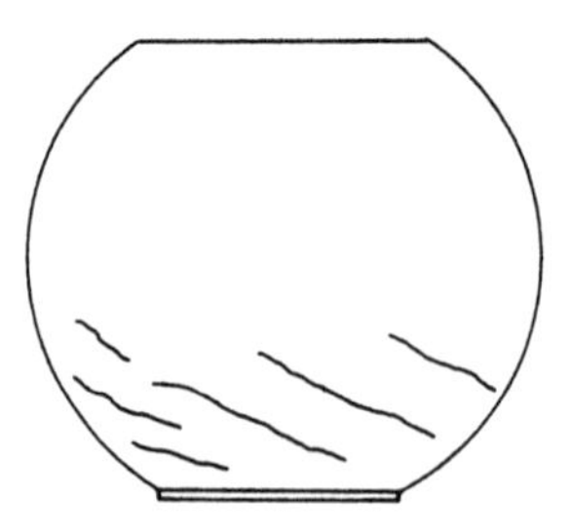

Fig. 9–4 Spiral cracks appear during the firing operation. These are caused by the twist of the clay wall during the throwing operation. This results in slippage of one plane of the clay with respect to an adjacent plane. The extensibility of the clay is exceeded in localized areas. These minute cracks do not heal, and even though they may not be visible prior to firing, they open up during the firing operation.

mation and taking care during the throwing operation not to exceed the extensibility of the clay.

PREHEAT CRACKS

This type of crack is characterized by remaining open, possibly either being filled with glaze or having the glaze pulled back away from it, depending on the glaze properties. It is a jagged fracture remaining open widest at the surface of the ware, and occurs upon heating in the temperature range of 550 to 650°C (1022 to 1202°F).

Two conditions are responsible for the development of preheat cracks:

1. A temperature gradient exists through the ware as the clay undergoes dehydroxylation. This is an endothermic reaction, so the center of the ware may become colder than the outside even though the kiln cycle may not appear to be excessively rapid.
2. The ware expands slightly during initial stages of firing, but upon completion of dehydroxylation shrinkage begins, resulting in a condition in which the outside begins to shrink while the interior is still expanding.

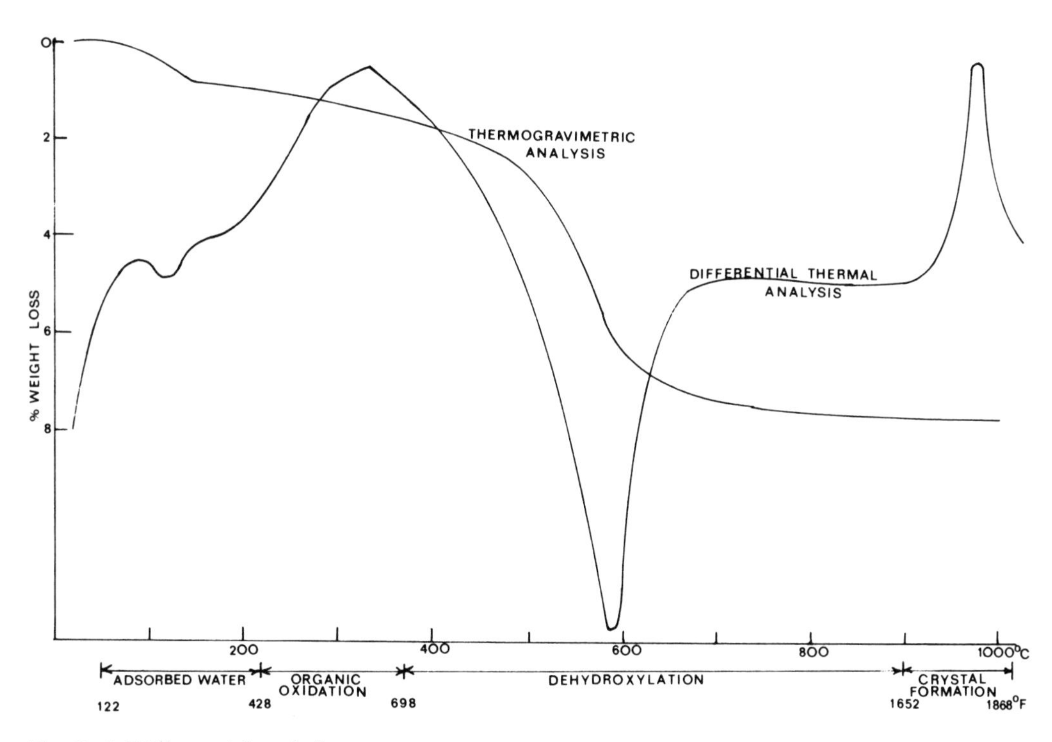

Fig. 9–5 Differential and thermogravimetric analyses of a porcelain insulator body.

Figure 9–5 shows the differential thermal analysis and thermogravimetric analysis run on a whiteware body containing approximately 30 percent clay. The analyses were run on small specimens weighing a few grams which had been ground finely and heated at a controlled linear rate of 20°C/min. (approximately 36°F/min.). Figure 9–6 shows the change in length of a small specimen of ware heated at a rate of about 5°C/min. (approximately 9°F/min.). The specimen expands until approximately 600°C (1112°F), the dehydroxylation temperature, and then begins to shrink.

Figure 9–7 shows that a piece of whiteware in the form of a cylinder, 3 in. in diameter by 6 in. in length, heated in an electric kiln with a differential thermocouple indicating temperature differ-

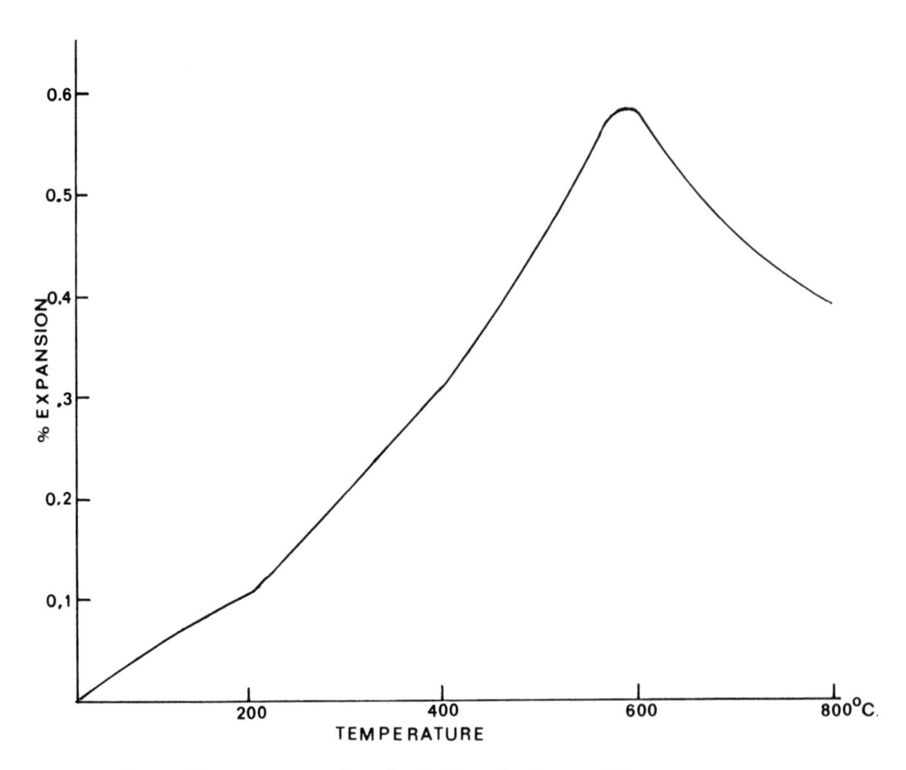

Fig. 9–6 Thermal expansion of unfired electrical procelain.

ence between the outside and the center of the piece, undergoes a series of thermal and physical conditions that could lead to cracking of the ware on the outside. The ware heated at a slow rate of only 0.8°C/min. (1.4°F/min.) shows a temperature gradient in the ware from the outside at 600°C (1112°F) while the inside is only 550°C (1022°F). This is the temperature at which the outside of the ware begins shrinking while the inside is still expanding. Figure 9–8 shows a plot of the inside of the ware versus the outside, or kiln, temperature at various times during the critical temperature range when preheat cracks form. Figure 9–9 shows a plot of the linear change at the inside of the ware versus the outside of the ware at kiln temperatures that are critical for the formation of preheat cracks. In the temperature

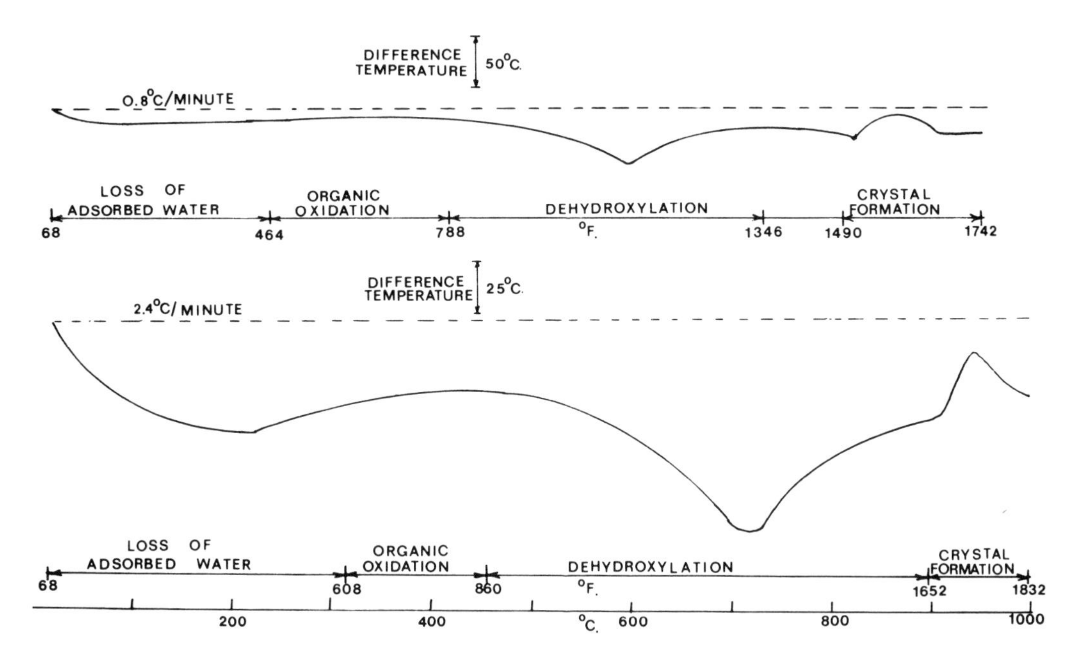

Fig. 9–7 Temperature difference between outside and center of 3 in. diameter electrical porcelain body heated at 0.8 and 2.4°C/min.

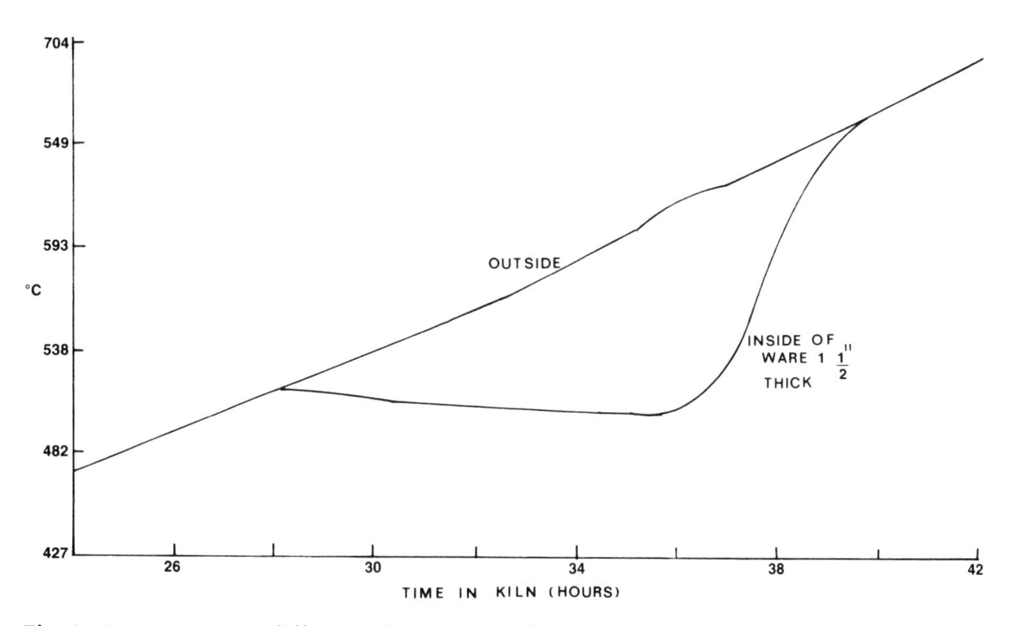

Fig. 9–8 Temperature difference between inside and outside of ware fired in periodic kiln.

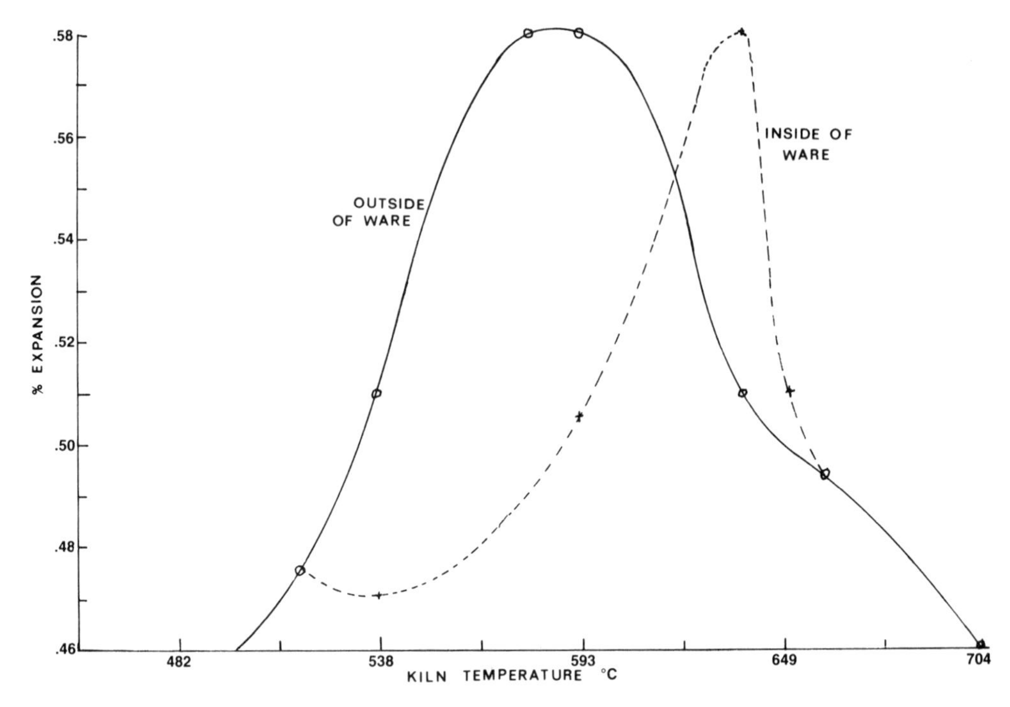

Fig. 9–9 Thermal expansion of outside and inside of ware fired in periodic kiln.

range during which the outside of the ware undergoes the major amount of expansion, cracks form in tension at the outside of the ware and progress inward. The cracks remain open during the remainder of the fire because there is no stress to close them.

A number of precautions may be observed to prevent preheat cracks:

1. During the critical temperature range of 550 to 650°C (1022 to 1202°F) the ware is being heated mainly by the convection currents of kiln gases because red heat has not yet been attained. Large amounts of excess air (more than 100%) in the combustion gases cause a more dynamic condition of the gases circulating about the ware and allows a lower temperature gradient to exist throughout the kiln and in the ware in the kiln.

2. Large volumes of water vapor emanate from the ware during dehydroxylation. This blankets the ware impeding equalization of temperature throughout the load of ware. Large amounts of excess air in the combustion gases help to remove the vapor from around the ware and the kiln.
3. Normally the kiln temperature is increased at a slow rate during the oxidation period of firing, and the rate is increased during the preheat stage. The kiln temperature is usually measured by crown thermocouples that do not represent the actual temperatures at the interior of large sections of ware. Extra time should be allowed at a slow rate of heating above 650°C (1202°F) if there is evidence that the interior of the ware is still cold and susceptible to preheat cracks.

DUNTING

Dunting occurs only after the ware has cooled to become an elastic solid. During the maturing range of firing, feldspathic glass is a viscous liquid, and, although the ware might deform or sag by prolonged heating, a thermal shock would have liffle effect upon it. However, as the ware is cooled to around 700°C (1292°F), the glass phase becomes more viscous and, at approximately this temperature, can be considered an elastic solid. Beginning at 600°C (1112°F) and continuing to about 400°C (752°F), the quartz in the ware undergoes a rapid linear shrinkage during and immediately following the transition from the β to the α form. A graph of the linear change of quartz during heating or cooling shows that approximately 80 percent of the shrinkage occurs in the 400 to 600°C (752 to 1112°F) range. Thus, if ware is cooled rapidly in this range and a temperature gradient exists in the ware from the outside to the inside, the rapid shrinking of the ware at the outside may cause a tension fracture that will close as it is cooled further because the ware has become elastic at this stage.

Some dunts have surfaces that are glassy smooth and appear fire polished. These cracks occur around the upper temperature range of 600°C (1112°F) where the glass is still very weak. Cracks that are irregular at the fracture occur at a lower temperature range of 400°C (752°F) where the glassy phase has attained a higher degree of strength.

Dunts may also occur upon heating bisque ware during a glaze fire in the range 400 to 600°C (752 to 1112°F), but in this case the cracks occur at the center of the ware which is cooler than the outside and is expanding more slowly.

Dunts may be avoided by eliminating cool drafts in the kiln during the slow cooling zone from 600 to 400°C (1112 to 752°F). Crown thermocouples may not be indicating the true condition of the ware in this zone, and extra precautions should be taken to avoid drafts. If the dunt is fire polished, drafts should be avoided in the 600°C (1112°F) temperature range. If the dunt fracture is irregular, care should be exercised in the 400°C (752°F) range.

Role of Cristobalite

Small amounts of cristobalite are present in many whiteware products and other fired clay bodies, and this poses some advantages as well as some problems. High-strength electrical porcelain that contains substantial amounts of cristobalite[8] is being produced, although the amounts must be carefully controlled so that the ware does not fracture from thermal shock at relatively low temperatures (Figure 9–10). Cristobalite undergoes a large nonlinear expansion when heated in the temperature range below 300°C (572°F), but the ware is less vulnerable to this range of temperature than the temperature range at which ware dunts (400–600°C or 752–1112°F) because of the large nonlinear expansion of quartz. The ware is less likely to be subjected to large thermal shock close to room temperature and also the glassy phase in the ware is much stronger at the relatively low temperature so that it may resist thermal shock better. In fact, this is probably the reason for the strengthening of the ware: the glassy phase surrounding the cristobalite is exposed to compressive strain (see Figure 8–3) that must be overcome before the glass fractures in tension.

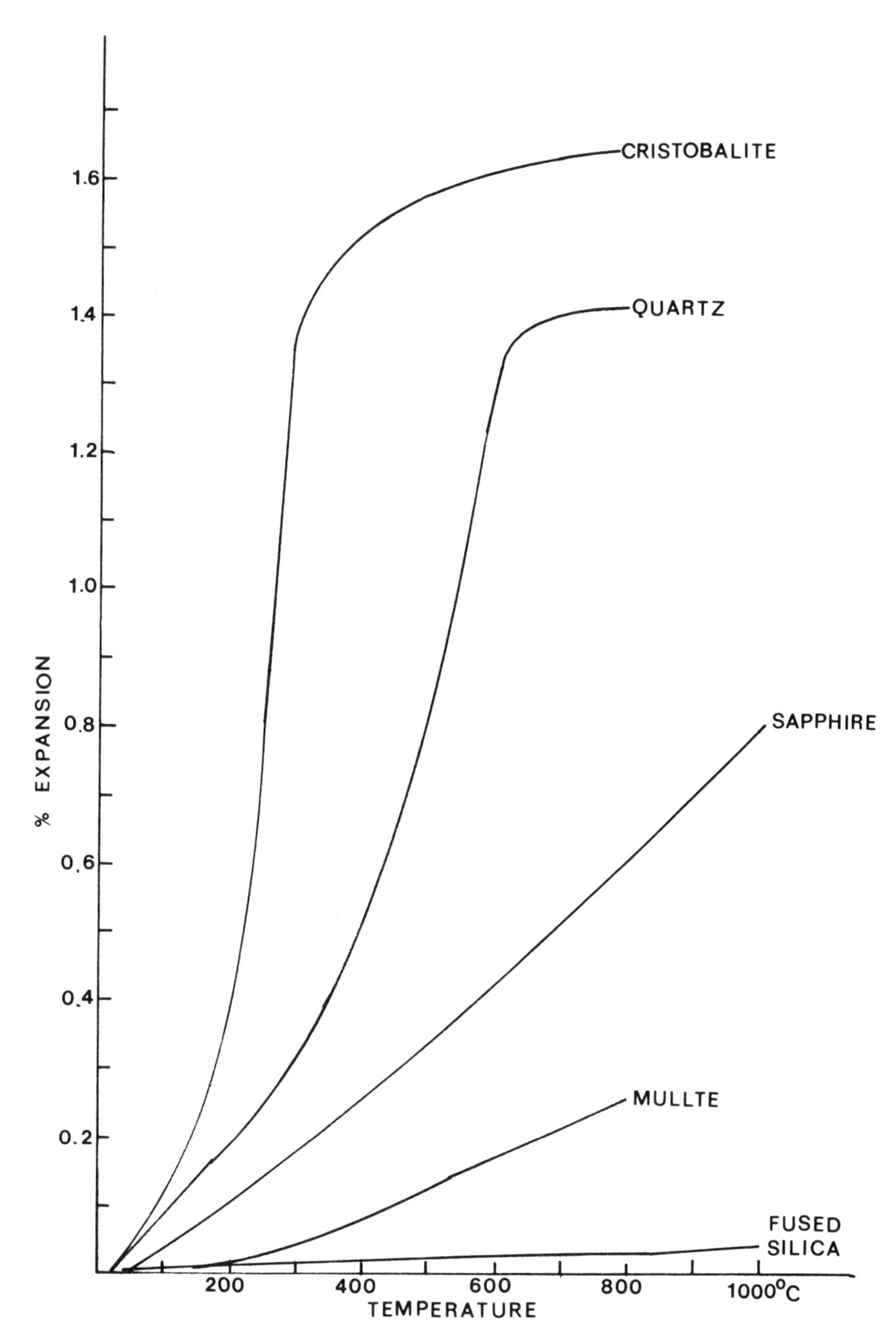

Fig. 9–10 Thermal expansion of fired whiteware body components.

Hotel china dinnerware is now being washed and dried at higher temperatures in order to observe stringent hygienic standards, and the ware manufacturers feel that some ware may have been broken because of thermal shock in the dishwashers. Therefore an ASTM test method[9] was devised to determine the thermal shock resistance of the ware from room temperature to 266°C (511°F). Hotel china that contains quartz has a tendency to fracture because of thermal shock in this temperature range, and in all probability this tendency may be related to the presence of about 1 percent cristobalite in the ware. Thermal shock will be discussed at length in Chapters 14 and 15.

Microscopic examination of whiteware does not show individual crystals of cristobalite but rather quartz crystals with a lacy coating of cristobalite. The development of cristobalite during firing of whiteware is not thoroughly understood, but in all probability the quartz does not change directly to cristobalite but rather passes through a transition phase.[10,11] A relatively high firing temperature is required to change quartz to this transition phase, but at normal whiteware firing temperatures some elements, such as calcium and possibly iron, may accelerate this change. The cristobalite appears to crystallize from this transition phase during cooling in the range 1100 to 1000°C (2012 to 1832°F). Rapid cooling in this range seems to prevent the crystallization of cristobalite and slow cooling in this range seems to accelerate the development of cristobalite. Trace amounts of alumina[12] and potassium[13] may have a catalytic effect upon the development of cristobalite. The particle size (surface area) of the quartz also is a major factor in the development of cristobalite[14] because a very fine quartz particle may be more easily induced to form cristobalite.

Kiln Setting

Ware set on refractory shelves during firing does not expand or contract at the same rate as the shelves. This causes some of the ware to crack. Round-grained glass sand between the ware and the shelves alleviates this problem somewhat, but the sand reacts slightly with the ware causing it to stick to the bottoms of pieces and the loose sand often poses a problem by sifting into the cracks of the kiln refractory. Shrinkage rings are used with large and heavy pieces that might have a tendency to drag on the shelves during shrinkage. These shrinkage rings are sections of body having the same composition and moisture content as the ware. These rings are usually about 1 in. thick and are placed between the ware and the kiln shelf. The rings usually crack but do prevent the ware from cracking.

The interface between the ware and the refractory shelf may show slight reaction at the maturing temperature. This causes the ware to sinter to the shelf and stick. Inasmuch as the cooling shrinkage between the ware and the shelf is different, a small section of the bottom of the ware may be pulled off leaving it stuck to the shelf. This is sometimes called plucking. To prevent this interreaction between the ware and the shelf, a kiln wash is painted on the refractory shelves. A mixture of 50 percent Florida kaolin with 50 percent alumina hydrate mixed with water to a creamy consistency makes a kiln wash that protects the ware from sticking and also precludes the use of setting sand which is messy.

REFERENCES

1. Fay, G. C. and Donahey, J. W., *North American Combustion Handbook,* N. A. Mfg. Co., Cleveland, Ohio, 1965.
2. ———, "Fuel Conservation in Kiln Firing-An Overview," *Bull. Amr. Ceram. Soc.,* 56, 8: 720–23 (1977).
3. ———, "Tempest Burners, High Velocity for Recirculation," N. Am. Mfg. Co. Bulletin 44.42, November, 1969.
4. Brownell, W. E., "Combustion and Kiln Atmosphere," *Firing Ceramics,* G. A. Kirkendale, ed., SUNY College of Ceramics, Alfred, N.Y., 1968.
5. Badger, A. E., "Effect of Various Gaseous Atmospheres on the Vitrification of Ceramic Bodies," *J. Am. Ceram. Soc.,* 16: 107 (1933).

6. Norton, F. H., *Fine Ceramics, Technology and Applications,* p. 81, McGraw-Hill Book Co., Inc., N.Y., 1970.
7. Jarrett, E. T., "Operating and Maintaining Today's Modern Kilns," *Firing Ceramics,* G. A. Kirkendale, ed., SUNY College of Ceramics, Alfred, N.Y., 1968.
8. Yamamoto, N., et al., Method of Producing Porcelain Articles, U. S. Patent 3,459,567, August 5, 1969.
9. ———, Crazing Resistance of Fired Glazed Ceramic Whitewares by a Thermal Shock Method, ASTM Test Method C 554, 1977.
10. Chaklader, A.C.D. and Roberts, A. D., "Transformation of Quartz to Cristobalite," *J. Am. Ceram. Soc.,* 44, 1: 35–41 (1961).
11. Kuellmer, F. J. and Poe, T. I., "The Quartz-Cristobalite Transformation," *J. Am. Ceram. Soc.,* 47, 6: 311–12 (1964).
12. Chaklader, A.C.D., "Effect of Trace Al_2O_3 on Transformation of Quartz to Cristobalite," *J. Am. Ceram. Soc.,* 44, 4: 175–80 (1961).
13. Christensen, N. H., Cooper, A. R. and Rawal, B. S., "Kinetics of Dendritic Precipitation of Cristobalite from a Potassium Silicate Melt," *J. Am. Ceram. Soc.,* 56, 11: 557–61 (1973).
14. Chaklader, A.C.D., "Particle Size Dependence of the Quartz-Cristobalite Transformation," *Trans. Brit. Ceram. Soc.,* 63: 289–300 (1964).

CHAPTER 10

AIR POLLUTION

The present concern about air pollution at the national, state and local levels makes it imperative for the individual potter to become aware of its causes and cures. Concern is no longer limited to large industrial establishments. Cases are already on record of small individual potters being cited for violation of air pollution ordinances.

Because of the variations in such ordinances at the state and local levels, it is impossible to discuss them in detail. But we will describe the types of pollution caused by potters' kilns and how they may be minimized. Because the types of pollution vary considerably with the type of fuel used, the discussion will be limited to kilns fired by natural gas and fuel oil.

As indicated in Chapter 9, the perfect combustion of a hydrocarbon fuel in air results in the formation of carbon dioxide, CO_2, and water vapor, H_2O. These gases are not considered to be air pollutants because they are completely harmless. However, when combustion is imperfect, as when not enough air is used to oxidize the fuel completely, carbon monoxide, CO, is formed, and, in extreme cases, soot or carbon particles are formed, producing smoke. These products of improper combustion are pollutants, and steps must be taken to minimize their emission into the atmosphere.

Most fuels contain varying amounts of sulfur. Coal is the worst offender because it may contain from 1 to 3 percent sulfur.[1] A No. 2 fuel oil may contain 1 percent sulfur, while natural gas contains only insignificant amounts. When these fuels are burned, sulfur dioxide, SO_2, and sulfur trioxide, SO_3, are formed. Both are considered serious and irritating air pollutants.

At high temperatures (above 1538°C or 2800°F) the oxides of nitrogen, nitric oxide, NO, and nitrogen dioxide, NO_2 are formed.

The source of the nitrogen is the air used for combustion. Small amounts of these gases are always generated at high temperatures.

In addition to the above inorganic gases, incomplete combustion may result in the formation of organic compounds, such as aldehydes, organic acids and methyl alcohol, particularly when the walls of the firing chamber are cool. The formaldehyde odor is often discernible at the start of the firing operation.

We will now discuss the formation of these various pollutants in more detail and make suggestions for their minimization or elimination.

Carbon Monoxide

The combustion of a hydrocarbon fuel involves the following chemical reaction. In the case of natural gas, we will assume the major constituent to be methane, CH_4.

$$CH_4 + 2O_2 \rightarrow CO_2 + 2H_2O \text{ (ideal combustion)}$$

If there is insufficient oxygen to oxidize the fuel completely, regardless of its type, the following reaction will result:

$$3CH_4 + 5O_2 \rightarrow CO_2 + 2CO + 6H_2O$$

Even though the total supply of oxygen may be sufficient for combustion of the fuel, insufficient mixing of the air and gas may result in imperfect combustion and the generation of CO.

The potter often wants to fire the ware under reducing conditions. This kind of firing calls for a fuel:air ratio higher than that which provides proper combustion and results in the formation of CO. In extreme cases it may also result in the formation of carbon particles (soot) that will be emitted from the stack in the form of smoke. Several steps may be taken to minimize formation of these products even while maintaining reducing conditions in the kiln chamber.

1. The potter is prone to overdo the reduction process. By cutting the air supply to the burners to a minimum and getting deep yellow flames to emanate from every crack and port in the kiln, the potter feels assured that good reducing conditions are present. This is certainly true but completely unnecessary. The presence of the CO gas in the firing chamber is responsible for the reduction process. In a gas-or-oil-fired kiln, a large volume of gas is moving over and around the ware. It is not necessary to have a high concentration of CO present to obtain reduction. Because CO is actually a fuel and will burn if any oxygen is present, its presence in small amounts in the firing chamber guarantees that there is no free oxygen present and that good reducing conditions prevail. The presence of large volumes of luminous yellow flames is completely unnecessary. The luminous flame contains unburned carbon particles (soot)—and carbon plays no role in the reduction process. Its presence merely attests to an improper control of the fuel:air ratio. Reduction can and should take place with no smoke visible from the stack.
2. In a ceramic kiln, it is a simple matter to introduce secondary air into the gas effluent stream after its passage through the firing chamber into the stack. This will provide additional oxygen for the burning of any CO to form CO_2, and also complete the combustion of any unburned carbon particles before they leave the stack. For these reactions to take place by simple air injection, the gas temperatures must be in the 538 to 816°C (1000 to 1500°F) region, and the gases must contain enough combustible material to develop a flame in the presence of the auxiliary air that is introduced. If these two conditions cannot be satisfied, simple secondary air injection cannot solve the problem.
3. Thermal oxidation of the contaminants may be accomplished by exposure to a fuel:air flame and by holding the contaminants in a chamber long enough to complete the combustion. The contaminated gases are raised to between 538 and 816°C (1000 and 1500°F) by mixing with the flame.[2] But

in a ceramic kiln being fired under reducing conditions, there would not be enough oxygen in the effluent gases to support combustion. It would, therefore, be necessary to use a premix type of afterburner in which the gas and air, in proper amounts, are combined and fed into the burner. The excess air in the burner supplements the oxygen in the stack gases to produce a total oxygen content sufficient to support combustion of the unoxidized gases, mainly CO. A typical commercial afterburner is illustrated in Figure 10–1, and several commercial designs are available.[3] This type of afterburner is capable of caring for unburned gases as well as small amounts of unburned carbon particles. If large amounts of soot or unburned carbon particles are generated, despite the fact that they are not necessary for the reducing process, further steps must be taken to prevent their emission from the stack. This will be discussed in the section on smoke.

Sulfur Oxides

Air contaminants classified as sulfur oxides consist essentially of two compounds, colorless and invisible sulfur dioxide, SO_2, and hygroscopic sulfur trioxide, SO_3. The source of both is the combustion of fuels that contain sulfur.

The limits of sulfur content in fuel oils are given by Magill et al.[4] as follows:

Grade of Fuel Oil	Sulfur Max. %
No. 1	0.5
No. 2	1.0
Nos. 4, 5, 6	No limit

Natural gas contains up to 100 grains per 1000 ft.[3] while the eastern bituminous coals vary from 0.3 to 5.0 percent sulfur.[4] Great progress is now being made in upgrading coals by purification processes at the mine, as well as by the removal of sulfur from the fuel oils. The

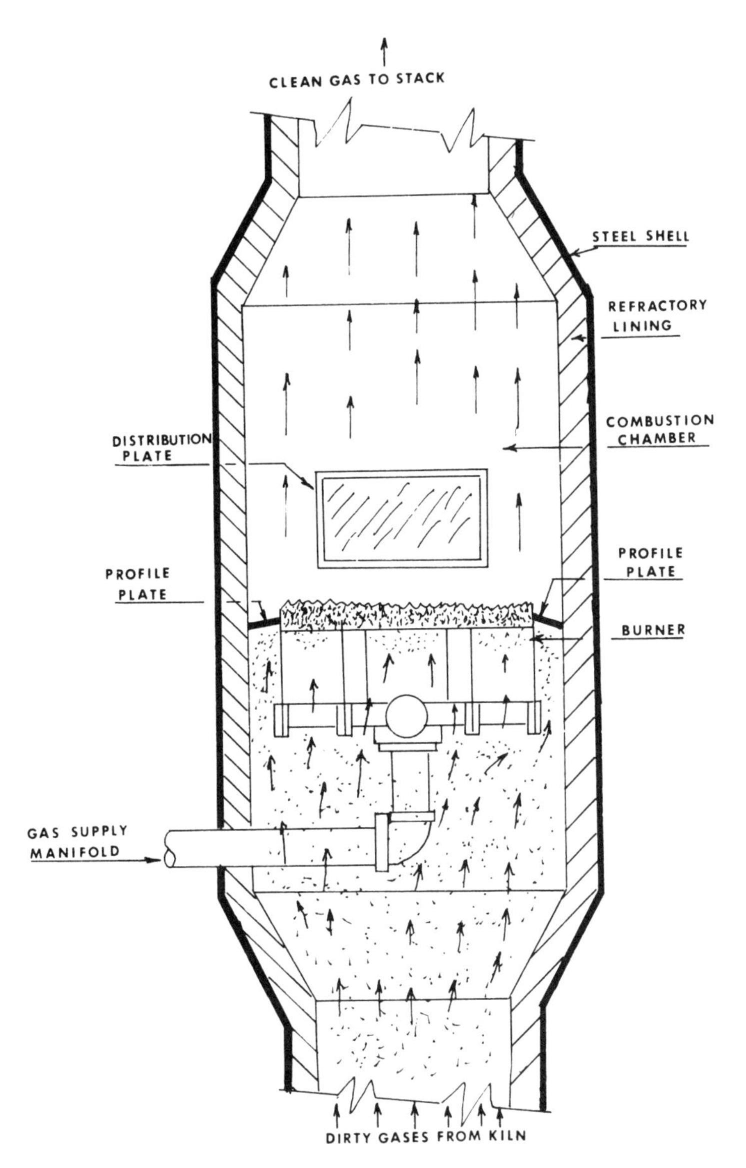

Fig. 10–1 A typical commercially available afterburner. Such units are capable of completing the combustion of gases such as carbon monoxide, organic gases and unburned carbon soot particles. Several types are available.

natural gas fuels are the cleanest and have very low sulfur contents. The potter should know what the sulfur content of his or her fuel is. It is now possible to specify low sulfur coal and oil which reduces the emission of SO_2 and SO_3 at the source.

Another source of sulfur oxides is the presence of sulfides or sulfates in the raw materials being fired. As these are decomposed or oxidized they will form SO_2 and SO_3. In most of the refined clays, the amounts present are insignificant but in some of the unrefined surface clays appreciable amounts may cause the formation of considerable amounts of these gases.

The gaseous sulfur oxides are very toxic. Both SO_2 and SO_3 can produce lung injury at small concentrations of 5 to 10 ppm.[1] Each can combine with water vapor to form acid aerosols which damage vegetation, metals and fabrics. At concentrations of 5 ppm SO_2 is irritating to the eyes and respiratory system.

The presence of SO_2 and SO_3 is often indicated by the formation of a blue haze at or near the top of flue stacks. This haze is caused by the oxidation of sulfur to sulfur trioxide. One to five percent of the total sulfur content of the fuel is converted to SO_3 under normal conditions of excess air combustion.[4] The SO_3 is then hydrated by water vapor in the stack or atmosphere gases to form aerosols of sulfuric acid, H_2SO_4. Their particle size is such that they appear blue against a dark background.

The colorless SO_2 is gradually oxidized to SO_3 after emission, and then it too forms sulfuric acid. This is eventually converted to sulfates that settle or are washed out of the atmosphere by rainfall.

The control of SO_2 and SO_3 formation involves many factors. Reduction of excess air to almost stoichiometric proportions(the minimum amount of air required for complete combustion) reduces the SO_3 produced by oil-fired furnaces. In such procedures, precise control of air:fuel ratios and good mixing is necessary to prevent smoke formation.

Some techniques used to remove SO_2 and SO_3 from flue gases involve wet scrubbing of the effluent gases, passing the flue gases over a catalyst to form H_2SO_4, reaction of SO_2 with metal oxides to produce sulfur, and reaction of flue gases with limestone to form calcium sulfate, which is then disposed of.[5] All of these methods involve expensive equipment beyond the means of most potters.

Control of sulfur oxides is best accomplished by the use of low sulfur fuels, minimization of excess air consistent with complete combustion, use of a high stack to dissipate effluent gases and the use of clays containing minimum amounts of sulfides and sulfates.

Nitrogen Oxides

Any combustion process that produces high temperatures in the presence of atmospheric nitrogen and oxygen will produce nitrogen oxides.

Several nitrogen oxides may result. Nitric oxide, NO, and nitrogen dioxide, NO_2, or its equilibrium product N_2O_4, are considered to be of significance in their contribution to air pollution. Stern[6] indicates that in the Los Angeles area, measurements on emissions of nitrogen oxides from small stationary sources showed an average of 43 ppm in incinerators, 70 ppm in gas-fired boilers and 250 ppm in oil-fired boilers.

The amounts of nitrogen oxides emitted may be small, but they play an important role in smog development. Nitrogen dioxide decomposes in light, forming NO and atomic oxygen. The atomic oxygen reacts with the molecular oxygen to form ozone, $O + O_2 \rightarrow O_3$, and the NO reacts with hydrocarbon groups in polluted air to form more NO_2. The cycle repeats itself, and, thus, a small amount of NO_2 can result in the formation of large concentrations of ozone and oxidized hydrocarbons, both of which are poisonous to plant and animal life.

The concentrations of NO emitted are affected by the maximum temperatures attained, rate of cooling of the combustion gases, furnace design and, to a lesser, extent, by the fuel used.

Stern[6] shows that the amount of NO generated is a direct function of the furnace wall temperature. This relationship is shown in Table 10–1.

These data indicate that the potter need not be too concerned about nitrogen oxide formation. Extrapolation of the data indicates that little, if any, would be formed at or below 1427°C (2600°F).[7] A survey by Chass and George[8] shows that gas-fired ceramic kilns produce 3 to 66 ppm of NO_2, while an oil-fired kiln produces 20 to 27 ppm. The temperature of operation of these kilns at the time of the sampling was not given.

Slow cooling in the exhaust passage or stack allows the nitrogen oxides formed to decompose to oxygen and nitrogen. Rapid cooling fixes the nitrogen oxides in the exhaust gases. A stack of sufficient height to provide slow cooling aids in the natural decomposition and dissipation of these products.

Uniform distribution of the flame and elimination of hot spots also aid in the minimization of these oxides. In ceramic kilns where uniform temperature distribution is of primary concern for other reasons, this suggestion seems superfluous. But it is a point worth making in regard to power plant boilers where high flame tempera-

TABLE 10–1
EFFECT OF FURNACE WALL TEMPERATURE ON % NO GENERATED

Furnace Wall Temperature, °C[a]	Yield of NO, %
1538	0.26
1649	0.41
1760	0.77
1871	1.30
1982	1.55
2093	1.75

[a]Flame temperature estimated to be 149°C above wall temperature.

tures and rapid transfer of heat through a tube or wall is desirable for efficient operation.

It has been shown that blue flames produce NO in the amount 50 ppm, while yellow flames produce 50 to 100 ppm. This indicates that sufficient air for complete combustion of the fuel reduces the formation of NO.

Organic Gases

The hydrocarbon derivatives or oxygenates, associated with the combustion process are products of incomplete combustion, and include aldehydes, ketones, alcohols and organic acids. They tend to react with NO_2 and produce plant damage and eye irritation.

When an atmospheric burner is supplied with ample air, combustion is complete over a wide range of operating conditions. When the flame is enclosed in a sufficiently large chamber so that air flow or the natural flame is not impeded by the chamber size, combustion remains complete. But if the flame is restricted by the chamber walls, or if the walls are cooler than the flame, incomplete combustion will result and large amounts of carbon monoxide, aldehydes and other organic gases will be produced.

We have all experienced the strong odor of formaldehyde at the start of the firing operation. At this time the walls of the kiln are cold and often only one of the several burners will be started with an excessively long flame. When the flame reaches the kiln wall, the gases are quenched to a temperature below the ignition point and incomplete combustion results. The obvious remedy is to limit the flame so that it is short (less gas and more air). It is better to have short flames from several burners than a long flame from one burner.

Surveys show that ceramic kilns emit approximately 3 ppm of aldehydes,[8] not a serious level and much lower than other types of industrial heating operations. Only when combustion is hindered and cold walls quench the combustible gases are these pollutants at a serious level.

Smoke

Smoke is a submicron particle aerosol that comes from a combustion source and that obscures vision. Smoke usually contains very small amounts of particulate matter by weight but because of its light-scattering properties in the 0.3 to 0.5 μ range it may appear as an impenetrable mass. On the other hand, a stack emitting a much greater weight of larger particles (100 μ) may appear clear because they do not scatter light. Nevertheless, dense smoke fumes are undesirable in populated areas.

With all the knowledge about combustion theory and practices, no chimney should emit black smoke. All fuels can be burned smokelessly and efficiently.[4]

The basic causes of smoke emission stem from operation, maintenance, design and installation. Very little skill is required to burn natural gas without smoke; control of the air:gas ratio is all that is necessary. On the other hand, considerable skill is necessary to handle highly volatile coal without smoke, and, in certain cases, rather sophisticated combustion devices may be necessary.

The following principles cover smoke-free combustion of all types of fuels:[4]

1. Proper air:fuel ratio.
2. Adequate mixing of air and fuel at the proper time (especially combustible gases vaporized from the fuel).
3. Sufficient ignition temperatures on combustible gases.
4. Adequate furnace volume to allow time for burning.
5. Proper firebox setting height (adequate height above grate).

The following are often called the three "Ts" of combustion:

Temperature: 649 to 760°C (1200 to 1400°F)
Turbulence: As much as possible
Time: ¼ to ½ to sec.
plus
Oxygen 16.2%

There is no problem in smokeless firing with oil or natural gas. The steps necessary for smokeless firing with coal and wood are given by Magill et al.[4] Smokeless firing requires sufficient oxygen for complete combustion. This in turn requires some degree of control and duplication of firing conditions. Because of insufficient or imperfect mixing of air and fuel it is usually impractical to fire a fuel with only the theoretical (stoichiometric) amount of air. Excess air is necessary to assure complete combustion. Consequently, some air is unused in the combustion process. For each type of fuel and associated burning conditions there is a relationship between the percentage of excess air and unburned fuel leaving the furnace. For gas firing, excess air may be maintained at 25 percent without producing unburned products. Natural gas requires 9.8 $ft.^3$ of air per cubic foot of gas for proper combustion. Table 10–2 lists the stack effluents resulting from oil and natural gas firing.

These conditions are not applicable in reduction firing which requires incomplete combustion and leaves CO and possibly some unburned carbon particles. Methods of eliminating these products before emission from the stack have been discussed earlier in this chapter.

Free Crystalline Silica

The OSHA regulations concerning the occupational hazard of free silica in airborne respirable dust, has made ceramic workers aware of the dangers of silicosis.[9] Both quartz and cristobalite may be dan-

TABLE 10–2

STACK EFFLUENTS RESULTING FROM OIL AND NATURAL GAS FIRING OF CERAMIC POTTERY KILNS

Effluents	Formation Conditions	Recognition Methods	Corrective Measures
CO_2	Normal complete combustion of hydrocarbon fuels.	Colorless, odorless, nontoxic. Not considered a pollutant.	None necessary
H_2O	Normal complete combustion of hydrocarbon fuels. Water removal from clays in 450–600°C range.	Steam from stack, not considered a pollutant.	None necessary
CO	Incomplete combustion of fuel, insufficient air and/or quench of combustion gases, insufficient mixing or excessive reduction.	Colorless and odorless. Commercial indicators and recorders available. Measure excess air and maintain 25% minimum.	Adjust air:fuel ratio to provide more excess air. Install afterburner in stack to care for emission during reduction.

SO_2 SO_3	Oxidation of sulfur content of fuel.	Blue haze from stack. Irritation of eyes, nose and throat.	Specify low-sulfur fuel. If this is not possible and excessive amounts persist with minimum excess air, water scrubbing equipment may be necessary.
NO NO_2	Oxidation of atmospheric nitrogen in high-temperature flame.	Usually present in such small amounts that analysis by gas chromatography or spectroscopy is necessary.	Corrective measures usually not necessary because of small amounts generated at pottery kiln temperatures. Uniform temperature distribution minimizes generation of NO and NO_2.
Aldehydes	Restriction of flame by cold furnace walls resulting in quenching of combusting gases below ignition point.	Formaldehyde or alcohol odor. Eye, nose and throat irritation.	Fire cold kiln with several small burner flames with excess air rather than with one long flame. Avoid reducing conditions in cold kiln.
Smoke carbon	Insufficient air, improper mixing, insufficient furnace volume, excessive reduction.	Obvious smoke emission from stack.	Adjust air:fuel ratio, avoid overreduction, provide secondary air, install afterburner.

gerous in airborne dusts, provided the particles are in the size range 1 to 10 μ. Larger particles are too large to be trapped in lung passages, and smaller particles are usually passed off from the body in fluids. The trapped silica particles form a toxic fibrogenic enzyme that is released into the lungs decreasing the pulmonary function.

Dusts in many ceramic plants in this country and abroad[10] have been found to have airborne concentrations of dust above the "threshold limit value" which is 0.1 mg/m.^{3} of respirable-size quartz in the air samples taken from a device hung about the neck of a worker. The dust samples are usually time-weighted averages for an eight-hour period and must be analyzed by X-ray diffraction methods for the amount of crystalline silica present. Investigations are continuing on methods for reducing silica dusts in the air and preventing exposure of the workers to these dusts. Better ventilation around fettling operations and cleaner protective clothing has helped to reduce this exposure.

REFERENCES

1. Danielson, J. A., ed., *Air Pollution Engineering Manual,* U.S. Dept. of Health, Education and Welfare, Cincinnati, Ohio, 1967.
2. Waitkus, J., "Oxidation and Combustion Means to Air Pollution Control of Gaseous Emission," presented at a course on Principles and Practices of Air Pollution Control by the Office of Manpower Development of the U.S. Dept. of Health, Education and Welfare, Ft. Meade, Md., March 4, 1970.
3. ———., "Waste Heat Recovery and Air Pollution Control, How and Why," Seminar on Modern Techniques in Firing Ceramics, Ohio State University, June 12, 1969.
4. Magill, P. L., Holden, F. R. and Ackley, C., *Air Pollution Handbook,* McGraw-Hill Book Co., Inc., N.Y., 1956.
5. Spaite, P. W., "Reduction of Ambient Air Concentrations of Sulfur Oxides—Present and Future Prospects," Proceedings, 3rd National Conference on Air Pollution, Washington, D.C., December 12–14, 1966; U.S. Dept. of Health, Education and Welfare.

6. Stern, A. C., *Air Pollution,* vol. 3, *Sources of Air Pollution and Their Control,* Academic Press, N.Y., 1968.
7. Diehl, E. H., "Reduction of Emission of Oxides of Nitrogen—Present and Future Prospects," Proceedings, 3rd National Conference on Air Pollution, Washington, D.C., December 12–14, 1966; U.S. Dept. of Health, Education and Welfare.
8. Chass, R. L. and George, R. E., *J. Air Pollution Control Assoc.,* 10, 34–43 (1960).
9. Bhargava, O. P., Alexiou, A. S., Meilach, H. and Hines, W. G., "Sample Preparation in the Determination of Free Crystalline Silica in Respirable Dusts from Steel-Making Environments," American Laboratory, September, 27–31, 1979.
10. Bloor, W. A. and Eardley, R. E., "Environmental Conditions in Sanitary Whiteware Shops," *J. Brit. Ceram. Soc.,* 72, 2: 58–69 (1978).

CHAPTER 11

GLAZES

Introduction

The subject of glazes could appropriately be the first subject for discussion in any book on ceramics because once a suitable glaze composition has been developed any necessary modifications are usually made to the body in order to arrive at the appropriate glaze-body compatibility. The reason for this is that glazes are very sensitive to small changes in oxide composition resulting in unexpected effects on color, texture and firing range. A glaze has many functions, some of which have no measurable properties.

1. A glaze provides a smooth, decorative and an aesthetic coating.
2. It is often used to cover blemishes in the ware.
3. The glaze must present a cleanable surface which, in the case of electrical porcelain, allows the rain to clean pollution from the surface, thus preventing arc-over, or, in the case of dinnerware, a cleanable, sanitary surface for eating purposes.
4. It provides an impermeable surface to protect underglaze decorations or to prevent access of moisture to the body which might cause freeze-thaw damage in outdoor exposures.
5. It provides a protective coating that prevents mechanical damage, such as the scratching of dinnerware by eating implements, chemical attack by food acids and is resistant to other severe environments.
6. The glaze increases the strength of the ware, which helps to prevent chipping of dinnerware and also "gunshot" damage to electrical porcelain.
7. Glazes used in operating rooms must have a slight electrical conductivity in order to bleed away static electrical charges that otherwise might cause explosive conditions with the gases present. Electrically conductive glazes are used with

some electrical porcelain pieces to provide a heat source in order to prevent moisture condensation.

8. Optical effects which provide especially lustrous or glossy appearances are especially important in dinnerware. Some blue-colored bodies covered with a yellow glaze give an unusual green color which is most appealing.

Most of the measurable properties of glazes are unrelated to the actual function of the glaze, for example, thermal expansion, Young's modulus of elasticity, tensile strength, hardness, opacity, surface tension, index of refraction and the number of bubbles present. Many of the most important glaze characteristics are subjective in nature and difficult to assess with measurable results and, thus, to control technically.

There are two general types of glazes: raw and fritted. Raw glazes are usually used at higher temperatures—above Cone 6 for once-fired ware, including sanitary ware, electrical porcelain and porcelain artware. The advantage of raw glazes is that they use less expensive raw materials and their composition is easily modified. The high firing temperature produces a glaze resistant to mechanical and chemical attack.

Fritted glazes are normally used at lower temperatures, below Cone 6. Some of the materials used in the formulation of these glazes are water soluble, which necessitates the fritting or prereacting of these materials to render them insoluble prior to preparation of the glaze. Fritted glazes produce fewer flaws because they are prereacted. The disadvantages of fritted glazes include the difficulty in modifying the composition, high cost of raw materials, and, in many cases, limited source of supply. Fritted glazes are used on wall tile, dinnerware and artware.

Fritting is normally conducted on raw materials that contain lead to help reduce the lead solubility in the final glaze with possible toxic effects. It also helps reduce the volatilization of lead from the glaze during firing. The glaze raw materials containing boron are also fritted because boric acid and borax, the main sources, are water soluble.

Glaze Raw Materials

Some raw materials are more suitable for formulating glazes than others. Alkalies are normally introduced in the form of feldspar or soda ash, but nepheline syenite is much more active chemically in dissolving quartz and produces a glaze more free of quartz crystals, which is desirable. Lime or magnesia may be introduced in the form of whiting or dolomite, but these raw materials give off large quantities of carbon dioxide gas at the glazing temperatures. The use of wollastonite or talc as a source of calcium oxide or magnesium oxide reduces the bubbles in the glaze. Alumina dissolves more readily in the glaze if introduced as clay rather than as an aluminum oxide or hydrate. Opacifiers are important to cover ware blemishes, intensify colors and produce special color effects. They are expensive and must be used with care. The degree of opacity produced by a material is related to the following factors:

1. The difference between the index of refraction of the opacifier and the glaze glass.
2. Size and degree of dispersion of the opacifier. The degree of opacity increases with the number of times a beam of light is reflected during passage through the glaze. Some opacifier particles may be so small that they selectively reflect either bluish or greenish light waves to give an opal-like effect.
3. The opacifier must be refractory and nonreactive so that there is a sharp boundary between the opacifier particle and the glass. A solution rim impairs effectiveness.

Table 11–1 lists possible opacifiers and gives their indices of refraction, melting point and expansion coefficient. The refractive index should be as high above the glaze glass as possible, the melting point should be very high and the expansion coefficient should be slightly higher than the glaze glass. However, the cost of the materials is a very important factor that changes continuously.

TABLE 11–1
POSSIBLE GLAZE OPACIFIERS

Material	Refractive Index	Melting Point °C	Expansion Coefficient ($\times 10^{-6}$ in./in./°C)
Glaze glass	1.55		6.0–7.0
Alumina	1.76	2050	7.0–8.0
Mullite	1.64	1850	5.4
CeO_2	High	above 2600	8.6
SiO_2	1.54–1.59	1713	High
ThO_2	2.2	3050	9.0
SnO_2	2.0–2.1	1625	4.1
TiO_2	2.6	1830	8.8
ZrO_2	2.2	2715	6.5–7.5
$ZrO_2 \cdot SiO_2$	1.92–1.96	2430	4.1
Air	1.0	—	—

Zirconium silicates are normally used for opacifiers, even though the expansion coefficient is low and the price is high. Tin oxide is also used extensively. Air bubbles in a glass opacify it effectively but detract from most of the other properties. Quartz particles are almost invisible in the glass because of the close match of refractive indices, but the quartz particles produce serious flaws and cracks in the glass because of the large mismatch in the thermal expansion coefficients.

Florida kaolins are usually selected for suspending glaze constituents in water because they are fine grained, maintain desirable viscosity characteristics of the glaze slip and simplify the maintenance of proper application thickness. These clays have high dry strength, which is desirable for handling dry ware, low organic content, good adhesion characteristics and only medium shrinkage, which prevents the formation of drying cracks in the glaze.

Sometimes the dry glaze coating on the ware has insufficient hardness to resist abrasion during handling, which makes it desirable to use small amounts of other binders. Cornstarch, gum arabic and gum tragacanth are common additions. Methyl cellulose is used occasionally when additional dry glaze hardness is required. It also prevents too rapid drying or water absorption during the application of the glaze.

Formulating Glazes

Glazes have been formulated for thousands of years, but the first evidence of any scientific formulation came around 1900, when Seger[1] introduced the molar glaze formula. Instead of using weight percent of oxides, the number of moles is used. The oxides are divided into three categories, and the oxides in each category have similar effects on the properties of the glaze when they are added in equal molar quantities. The three categories are:

1. RO and R_2O oxides, which include the alkalies, Na_2O, K_2O, Li_2O, and alkaline earth oxides, CaO, MgO and BaO.
2. R_2O_3 oxides, usually Al_2O_3.
3. RO_2 oxides, usually SiO_2 but also includes B_2O_3, which acts in a similar manner.

In each case, the "R" stands for a radical, or some element to be combined with oxygen.

The following calculation indicates the method for changing the oxide percent into a molar glaze formula.

Group	Oxide	Weight %		Molecular Weight		Moles	Corrected Moles
R_2O	K_2O	10.9	÷	94.0	=	0.116	0.3
RO	CaO	15.2	÷	56.0	=	0.271	0.7
				Sum	=	0.387	1.0
R_2O_3	Al_2O_3	15.8	÷	102.0	=	0.155	0.4
RO_2	SiO_2	58.1	÷	60.0	=	0.968	2.5
	Total	100.0					

Each weight percent of oxide is divided by the molecular weight of that oxide to give the number of moles. This formulation of moles must then be corrected so that the sum of the RO plus R_2O oxides equals one. In this case the sum of these moles equals 0.387, so each of the molar amounts is divided by 0.387 to give the corrected molar formula. The ratio of alumina:silica is also important. In this case it is 0.4:2.5 or 1.0:6.25.

Based on this molar formula for a glaze, Seger listed the following rules for making the glaze less fusible or more refractory.

1. Increase the silica that actually lowers the basic fluxing content. The moles of silica are maintained in the 2.0 to 3.0 range because below this range the glaze tends to run and above this range the glaze devitrifies or crystallizes.
2. Substitute the less energetic fluxes; that is, replace PbO with K_2O, Na_2O, CaO or MgO.
3. Decrease the number of fluxes, using either Na_2O or K_2O instead of both.
4. Increase the alumina and silica keeping the ratio constant.
5. Decrease the B_2O_3 and substitute an equal number of moles of silica.

The logic for these rules is now understood, but at the time they were devised they were based upon close and methodical observations made in actual practice. Seger then listed rules for prevention of crazing of the glaze. This involves methods for increasing the crystalline silica in the body to give it a higher expansion or for increasing glassy silica in the glaze to give it a lower expansion. These rules are as follows:

Body:

1. Decrease clay content and substitute flint.
2. Decrease kaolin and substitute ball clay.
3. Decrease feldspar.
4. Use coarser grind of body.
5. Use softer bisque fire.

Glaze:

1. Decrease the fluxing oxides and add silica
2. Increase the B_2O_3 and decrease silica.

Glaze-Body Relationships

The concept fundamental to the understanding of glaze fit is that of compression and tension. If a piece is subjected to pressure on both sides as in Figure 11–1A, it will be placed under compression. If the applied force tends to stretch or elongate the piece, it will be placed in tension, Figure 11–1B. It is the nature of brittle materials, such as ceramics, to be strong in compression but weak in tension. It follows, therefore, that a stable, well-fitted glaze should be in compression.

The physical property of a body and glaze that determines the extent of compression or tension is their thermal expansion. This is the amount that a material expands when heated or contracts when cooled. The thermal expansion of a body or glaze depends, for the most part, on its composition and thermal history. Because the composition of the body and glaze can be independently varied, the potter has the opportunity to adjust the thermal expansion of one or the other to provide a proper glaze fit.

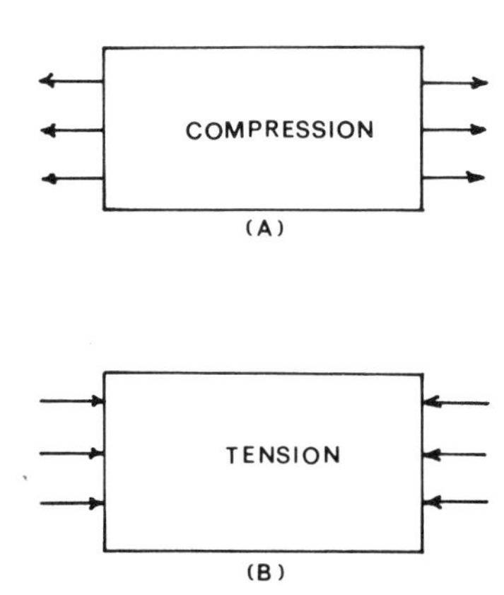

Fig. 11–1 Compression and tension resulting from applied force.

Figure 11–2 shows the relationships between the thermal expansion characteristics of a glaze and body and the compression-tension stresses developed on cooling.

In the example given, the glaze has a lower expansion than the body. On cooling from temperature, T_s (the temperature at which the glaze will first develop strain), the glaze would contract an amount, C_g, and the body would contract an amount, C_b, if they were separated. Because they are bonded, the glaze must be in compression and the body in tension after cooling. This is the most desirable and stable condition.

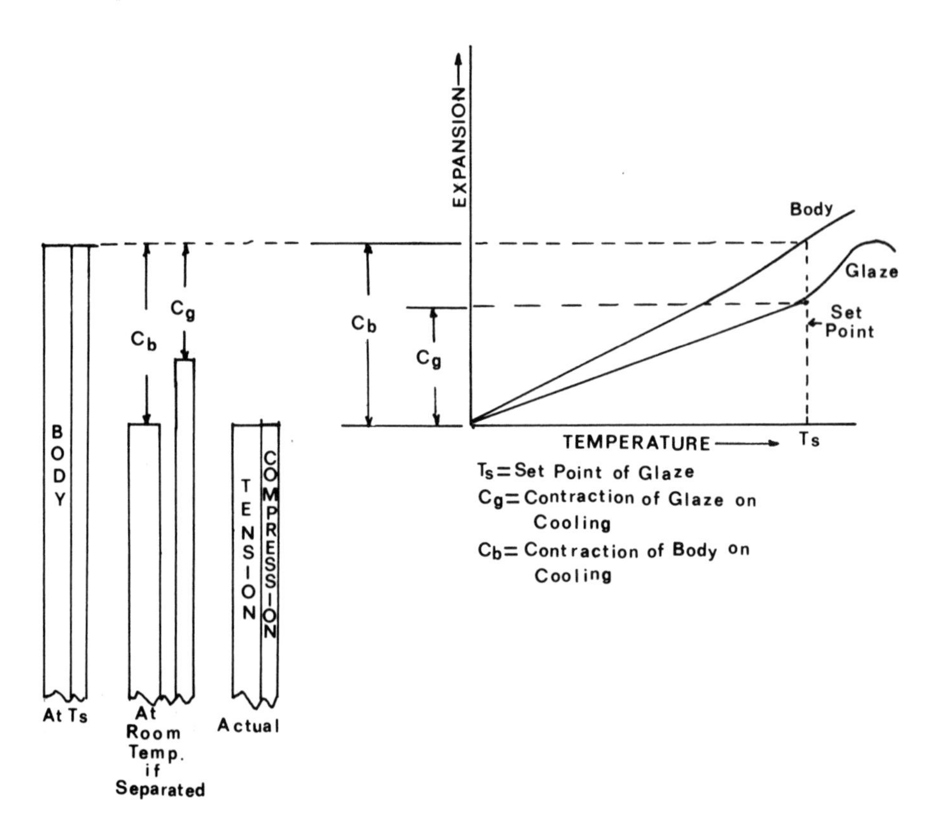

Fig. 11–2 Expansion-stress relationship in body and glaze resulting in the formation of compressive stress in glaze at room temperature.

The case of a glaze having a higher rate of expansion than the body is illustrated in Figure 11–3. On cooling from temperature, T_s, the glaze will contract an amount, C_g, providing they were separated. Because they are bonded, the glaze will be in tension and the body in compression, a condition conducive to tensile crazing. This type of craze pattern is illustrated in Figure 11–4 and will be discussed in detail in the section on glaze defects. The greater the tensile stresses in the glaze, the smaller the craze pattern. For purposes of decoration, the craze pattern may be controlled by manipulation of the expansion difference between the body and glaze.

Around 1930 a number of technical papers published in the *Journal of the American Ceramic Society* debated the relationship between the expansion of the glaze and the expansion of the body. Before

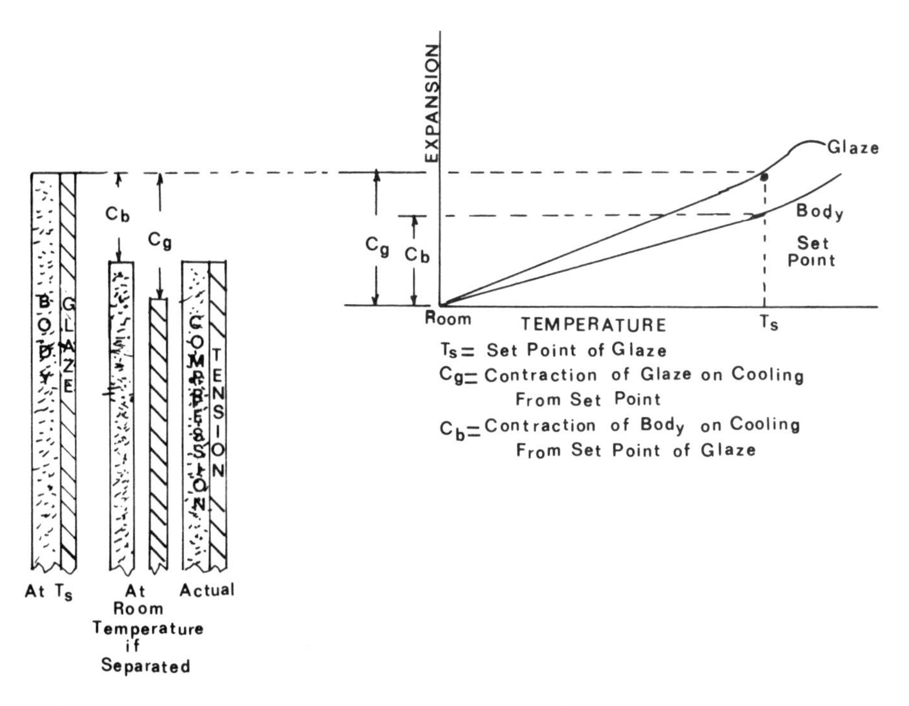

Fig. 11–3 Expansion-stress relationship in body and glaze resulting in the formation of tensile stress in glaze at room temperature.

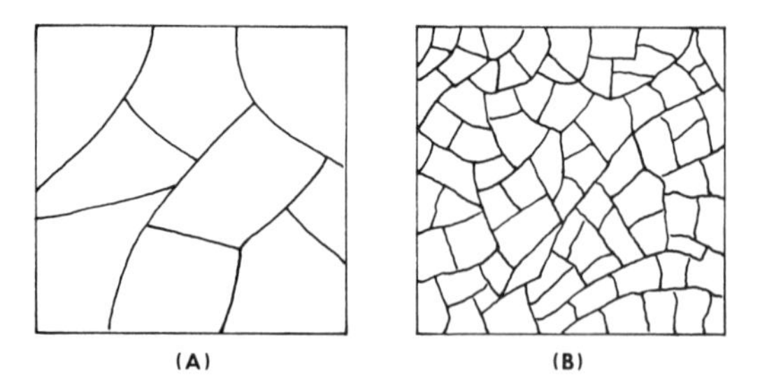

Fig. 11–4 Tensile crazing of a glaze. (A) Low stress, (B) high stress.

this time, thermal expansion measurements were difficult to perform and so this question had not arisen. The debate was ended with a technical paper by D. H. Rowland[2] and a subsequent patent[3] on a compression glaze. This stimulated a flurry of research by competitors to duplicate the compression glaze and attain the remarkable strengths that Rowland obtained with his glaze-body combination.

The wording of the patent is very technical and deals with electrical porcelain technology, but the major achievement was the understanding that the glaze expansion should be less than the body expansion by 5 to 15 percent, with the optimum being 10 percent lower expansion. This idea presents a challenging problem: How are glaze and body expansions measured and at what temperature are they to be compared?

A specimen of the body or glaze approximately ½ in. in diameter by 2 in. in length is heated slowly in a fused silica holder, and the difference between the specimen and the holder is measured by some accurate device, such as a dial gauge. The gauge reading must then be corrected for the amount the holder expands in order to give the actual expansion of the glaze or body. The observed expansions are then plotted on a graph over the range of temperatures measured. A typical graph is shown in Figure 11–5.

The first step in the comparison of the glaze expansion with the body expansion is to determine the set point temperature of the glaze, as shown in Figure 11–5. As noted in Chapter 8, this is the temper-

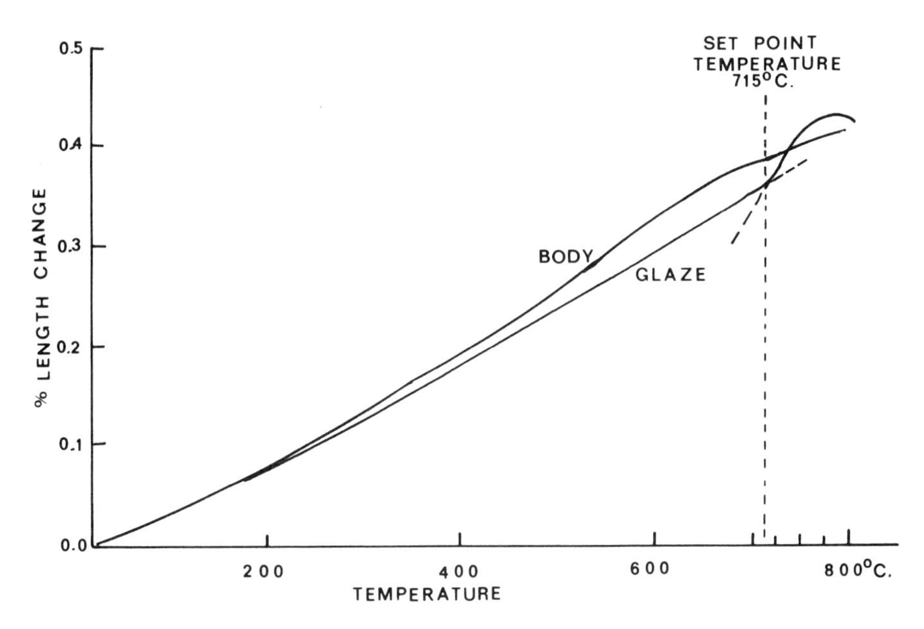

Fig. 11–5 Typical thermal expansion curve for body and glaze showing the graphic determination of the set point of the glaze.

ature at which the glaze attains a sufficiently high viscosity on cooling so that it acts like a rigid solid. In other words, it is the temperature below which the glaze becomes stressed as a result of the difference in shrinkage between the body and glaze. This temperature is determined by drawing tangents to the expansion curve as shown in the graph. In this case, the tangents intersect at a temperature of 715°C (1319°F) so this is the set point of the glaze. Actually the set point may be at some slightly higher temperature because the glaze usually reacts with the surface of the body so that the glaze composition at the interface is different from that of the glaze specimen tested.

The expansion of the glaze is measured in inches change in length per inch of specimen length, and at the set point the glaze has an expansion of 0.00365 in./in. This figure multiplied by 100 gives 0.365 percent expansion at the set point of 715°C (1319°F). The coefficient of expansion is the average expansion over the range of testing from 25 to 715°C or 0.00365/(715 − 25) = 5.29×10^{-6} in./in./°C.

The expansion of the body at 715°C is 0.00387 in./in. and the coefficient of thermal expansion is equal to 0.00387/(715 − 25) = 5.61×10^{-6} in./in./°C. The glaze applied to this body would have less shrinkage than the body and so would be in compression. The difference in shrinkage between the two would equal (0.00387 − 0.00365)/0.00387 × 100 or 5.7 percent. According to Rowland's patent, the difference should be slightly greater in order to attain the optimum strength.

The disadvantages of measuring glaze and body expansions using a dilatometer to control glaze-body fit are that the equipment is relatively expensive and difficult to use and that the actual set point temperature cannot be measured accurately because the interfacial reactions are unknown. Therefore, the dilatometer is used for partial control of glaze-body fit, and more reliance is placed on simpler control measures that test the glaze fired on the body, a more realistic situation.

One glaze fit test that is widely used is called the napkin ring test. A specimen of the body is formed into a cylindrical ring about ¼ in. thick by 1 in. long by about 4 in. in diameter. The glaze is applied to the outside circumference of the ring and then fired. Two points about ½ in. apart on one end of the ring are measured using a microscope micrometer. The ring is then cut with a diamond saw between the two points and the distance between the points is remeasured. A closing of the saw kerf by some predetermined distance indicates that the glaze was in compression.

A test that is more widely used and recognized is a simple transverse strength test. Glazed and unglazed specimens are prepared and then fired. The strength measured by breaking the specimens is an indication of the degree of glaze compression. The effect of glaze compression upon the strength of the body is more evident using ¼ in. diameter rods than by using ½ in. diameter rods. An increase in strength of the glazed rods by 50 percent as compared with the unglazed rods is considered excellent, although the stronger the unglazed strength, the higher the expected glazed strength. A poor glaze-body interface or a glaze that is in slight tension will produce glazed strengths that are lower than unglazed strengths.

Estimating Glaze Properties

Early work by Winkelmann and Schott[4] reported in 1894 showed that the thermal expansion of a glass may be estimated from its chemical composition. Recent studies on enamel glasses and glaze glasses show fairly close estimates of not only expansion, but also such properties as acid resistance, alkali resistance, specific gravity, refractive index, Young's modulus, gloss, surface hardness, flow at firing temperature and viscosity.[5–8] The accuracy of the estimate depends on the accuracy of the original test measurement. Thus, a test having a good accuracy, such as refractive index, will allow estimates with a confidence of 95 percent in a range of ± 1 percent. Values for thermal expansion may be made with a confidence of 95 percent in a range of ± 7 percent. This method for estimation allows an evaluation of a glaze without all of the time required for preparing specimens and testing them on expensive equipment. Coefficients of thermal expansion and fusion temperature are shown in Table 11–2. The thermal expansion measurements were made using an interferometer, and the fusion measurements were made on a special flow block, but all results were reported using the same equipment under standard conditions.

This illustrates the manner in which coefficients are used for estimating the properties of a glaze (Table 11–3). The moles of each of the oxides listed in the molar formula are multiplied by the coefficient for that oxide to give the contribution of that oxide to the property in question. The sum of all of the contributions of the oxides gives the estimated value of that property. The largest contribution to the expansion coefficient is made by CaO and the smallest by SiO_2.

TABLE 11–2

COEFFICIENTS FOR ESTIMATING PROPERTIES OF GLAZES[8]

	MOLAR COEFFICIENTS		*WEIGHT PERCENT COEFFICIENTS*	
Oxide	**Fusion Temperature in °F**	**Coefficient of Thermal Expansion**	**Fusion Temperature in °F**	**Coefficient of Thermal Expansion**
K_2O	1443.056	11.118	4.600	0.331
Na_2O	1238.420	12.332	−2.417	0.387
CaO	1625.083	8.553	21.762	0.148
MgO	2060.284	4.656	40.056	0.026
BaO	1740.896	9.158	18.193	0.129
SrO	1858.121	7.855	19.459	0.159
ZnO	1664.609	8.148	21.385	0.094
Al_2O_3	94.253	0.912	20.026	0.063
B_2O_3	−63.170	−0.556	11.670	0.031
SiO_2	43.090	−0.489	18.312	0.035
ZrO_2	−767.102	1.494	−8.254	0.099
F	89.034	0.768	38.750	−0.018
PbO	1171.973	7.452	9.925	0.083
TiO_2	−111.650	0.949	8.605	0.144

TABLE 11–3

ESTIMATING PROPERTIES OF GLAZE USING COEFFICIENTS

Oxide	Molecular Weight	Molar Formula		Expansion Coefficient of Oxide		Contribution to Expansion Coefficient of Glaze	Molar Formula		Fusion Coefficient of Oxide		Contribution to Fusion Temperature of Glaze (°F)
K_2O	94.0	0.067	×	11.118	=	0.7449	0.067	×	1443.056	=	96.68
Na_2O	62.0	0.015	×	12.332	=	0.1850	0.015	×	1238.420	=	18.58
CaO	56.0	0.346	×	8.553	=	2.9593	0.346	×	1625.083	=	562.28
MgO	40.0	0.124	×	4.656	=	0.5773	0.124	×	2060.284	=	255.48
ZnO	81.0	0.318	×	8.148	=	2.5911	0.318	×	1664.609	=	529.34
PbO	223.0	0.47	×	7.452	=	0.3502	0.047	×	1171.973	=	55.08
BaO	153.0	0.082	×	9.158	=	0.7510	0.082	×	1740.896	=	142.75
	Sum	0.999									
Al_2O_3	102.0	0.261	×	0.912	=	0.2380	0.261	×	94.253	=	24.60
SiO_2	60.0	2.363	×	−0.489	=	1.1555	2.363	×	43.090	=	101.82
						7.2413					1786.61°F

Estimates of properties: Expansion coefficient = 4.023 × 10^{-6} in./in./°F (7.241 × 10^{-6} in./in./°C)

Fusion temperature = 1786°F (974°C)

Measured properties: Expansion coefficient = 4.18 × 10^{-6} in./in./°F (7.53 × 10^{-6} in./in./°C)

Error of estimate: 3.8%

The largest contribution to the fusion temperature is also made by CaO and the smallest by Na_2O. Thus, it is easy to see that the coefficient of expansion might be reduced by increasing the SiO_2 slightly, but this also increases the fusion temperature. Note that the only oxide that reduces the expansion coefficient while reducing the fusion temperature is B_2O_3.

Glaze Adherence[9]

Three mechanisms have been proposed for the adherence of a glaze to a body:

1. Mechanical
   The interlocking of the glaze into irregularities on the surface of the body may be a partial explanation, but a roughening of the surface of the body does not appear to improve the glaze adherence.
2. Van der Waal's Forces
   These weak, intermolecular electrical forces are also partially responsible for adsorption, rely upon fixed adhesion sites, good wetting characteristics of the glaze and also good match in expansion of the body and glaze.
3. Dendrite Formation
   Reaction between the body and the glaze at the interface may cause the precipitation of dendritic or needle-like crystals from the glaze which may enhance adherence.

The formation of an interfacial layer is probably the most important of the mechanisms for glaze adherence and this accounts for a graded stress between the glaze and body which gives more freedom from crazing, a higher strength and an improved surface of the glaze. Huff examined the brown glaze used on electrical porcelain that is colored by oxides, including MnO_2. Dendritic crystals of $MnO_2 \cdot 2SiO_2$ formed at the interface, and the brown color of the glaze immediately surrounding these crystals was absent. The strength of the glazed specimens was 13,900 psi when MnO_2 was present but only 8,300 psi when the MnO_2 was removed from the glaze, even though the expansions of the glazes remained about the same.

Some glazes do not form much of an interfacial layer and the strength of the ware may be improved by applying a coating to the body before glazing. Smothers and Selsing determined that the coating should have a relatively low expansion coefficient of 3.2 to 4.0 $\times 10^{-6}$ in./in./°C. They proposed that the coating be made containing crystals of cordierite, spodumene or eucryptite. However, these coatings work with some glaze-body combinations but not with others.

Glaze Defects[10,11]

CRAZING

These cracks in the glaze are caused by tension in the glaze after firing and are usually produced by too low a quartz content in the body, overfiring, which reduces the quartz content, or a moisture expansion of the body after it has been placed in service. The autoclaving of ware usually shows whether the body is susceptible to moisture expansion and such ware can be corrected by firing to a higher temperature. The napkin ring test or transverse strength tests are the best methods of determining proper glaze fit.

CRAWLING

This is a surface tension effect that causes the glaze to pull away from some portion of the body leaving it bare. A number of causes may be responsible, including too thick a glaze application, too fine grinding of the glaze in the ball mill and ZnO in the glaze which has not been calcined or which has been calcined at too low a temperature. Shrinkage cracks in the glaze during drying may not heal during the early stages of firing, and sometimes the body may be dusty or dirty when the glaze was applied causing poor adherence. Using potters zinc (calcined ZnO) or PbO in the glaze usually cures crawling.

ENTRAPPED BUBBLES

There are a variety of categories of entrapped bubbles in the glaze surface from eggshell, which is a myriad of tiny bubbles in the glaze surface, to blisters, which are very large bubbles. Pinholes are also caused by large entrapped bubbles reaching the glaze surface and then bursting without healing.

Many of the reactions of glaze ingredients, or reactions of the glaze with the body, produce gas that accounts for bubbles. The glaze viscosity may be too high at the time of the reaction to allow the bubbles to fire from the glaze. Impurities, such as sulfur, in the kiln atmosphere may also produce bubbles.

Overfiring is more often the cause for entrapped bubbles than underfiring, but testing a glazed specimen in a thermal gradient furnace will indicate a firing range of the glaze which is bubble free. When overfiring is the cause for the bubbles, and a reduction in firing temperature is not desirable, the problem may be remedied by increasing the kaolin or alumina content and reducing the silica.

COLOR BLOWOUT

Some colors, such as blues and greens, are susceptible to color blowout when the color pigments have been calcined improperly or allowed to age for too long a time before using. Sometimes the color will cause the glaze to be completely blown from some portion of the ware; at other times the color just produces a large number of bubbles or pinholes at the colored area.

ORANGE PEEL

This is a spraying defect that causes the glazed surface to have the dimpled appearance of an orange. It is caused by too light a glaze application, which does not allow the glaze to flow out evenly; it may also be produced by a glaze that has a high viscosity at the firing temperature.

SCUMMING

The glaze may have a scummed appearance when sulfur is present in the kiln atmosphere or on the surface of the ware. Sulfates in the glaze may accumulate at the point on the ware that dries last. At times, an apparent scumming may be the result of devitrification of the glaze with small crystals of cristobalite present just under the glaze surface. A larger amount of excess air in the kiln atmosphere with a greater air turbulence usually reduces the scumming problem.

OPALESCENCE

Boric acid frits melted alone usually have an opalescence that is attributed to low alumina content. The addition of more alumina to the glaze usually improves this condition.

MOTTLING

A mottled appearance may be caused by the segregation of the opacifier of the glaze during the early stages of firing. Drying cracks in the glaze may cause a crawling in the early stages of firing which later smooth out at higher temperatures. A low-melting flux, such as a lead or a borate frit usually helps this condition.

Glaze Fit

The problem of glaze fit, and its deductive solution, may best be described by illustration. In the case that follows, the challenge was to reproduce a glaze whose method of creation remains unknown. The development of a proper glaze (and body) for the Blue Hippo is intriguing because the art itself has never been duplicated and yet the reproductions are very close to the original.

Approximately 3500 years ago, a blue glaze or glass was produced in the Middle East. This process became a fine art that was eventually lost and as yet has not been recovered.[12] Beads or figures may have been carved from some type of siliceous stone, coated with this blue glaze and buried with the dead to assist them in their passage into the Other World. Archeologists have discovered a number of the blue-glazed objects in Egypt, and they have become highly prized art objects.[13,14] Much speculation has resulted as to their method of fabrication.[15]

The Metropolitan Museum of Art was instrumental in unearthing the Tutankhamen tomb, and recently arranged for some of the most prized pieces of the collection to be shown at a number of museums in the United States and Canada. To help raise funds, reproductions of some of the collection were made for sale to the public. Making a reproduction that could be manufactured easily and sold at a nominal price posed a challenge with respect to one of the most prized items—a blue-glazed hippopotamus. No U. S. manufacturer could be found, and one foreign supplier encountered problems with glaze fit. The Met decided to manufacture the hippo in their own studios, but had to search for a person highly skilled in bodies and glazes to do the job. A graduate student in the College of Ceramics at Alfred University took the project as part of her graduate program.[16]

The original ware possessed a number of unusual qualities:

1. The body for the ware has been called faience because it is a soft type of ware fabricated from material consisting mainly of quartz, which might have been molded or carved from a stone.
2. The glaze film covers the entire body and lacks the usual "support marks" where the ware was set in a kiln.
3. Both the body and the glaze are very low in alumina, as shown by the chemical analysis of a similar glaze listed in *Ceramics and Man Through the Ages,* which was made available through the courtesy of the Metropolitan Museum of Art. Copper oxide is green in a glaze if alumina is present in substantial amounts in either the glaze or body.
4. Williamson of Penn State has theorized that the glaze was produced by burying the ware in a bed of powder contained in

a refractory saggar. When the ware was fired, alkali salts migrated through the powder bed causing the ware to become coated with a glass by a process similar to salt glazing. The powder bed sintered only slightly and shrank away from the ware. Although this process explains the absence of the support marks and the high value placed upon the figurines by the Egyptians, it does not present an easy method for manufacturing reproductions in an economical manner.

BLUE HIPPO GLAZE

The chemical analysis furnished for a blue glaze removed from an Egyptian art object was used to estimate the fusion temperature and expansion coefficient of the glaze. These estimates were used only as reference points because, in all probability, the glaze was applied by alkali migration through a powder bed and reaction with the body surface. Table 11–4 shows the calculation of the estimated properties and indicates that the fusion temperature is exceptionally low and the expansion coefficient very high. The next step in the development was to compound a number of glazes with approximately the same composition and test these for color, surface finish, fusion temperature and thermal expansion. One acceptable glaze was compounded and the estimated properties were compared with measured properties for the glaze (Table 11–5). The alumina content was kept low by using bentonite to suspend the glaze slip and by using glaze frits low in alumina. The thermal expansion graph for the glaze is shown in Figure 11–6.

BLUE HIPPO BODY

To make a fired body with good handling strength, high thermal expansion to fit the blue glaze and low alumina content to prevent the glaze from firing to a green color, a number of high-quartz bodies were prepared. Twenty-five percent clay was required for acceptable

TABLE 11–4

ESTIMATING PROPERTIES OF BLUE GLAZE FROM CHEMICAL ANALYSIS

Chemical Analysis			Fusion Coefficient of Oxide		Contribution to Fusion Coefficient of Glaze (°F)	Chemical Analysis		Expansion Coefficient of Oxide		Contribution to Expansion Coefficient of Glaze
SiO_2	65.10%	×	18.3	=	1191.3	65.10%	×	0.04	=	2.60
Fe_2O_3	1.21		?			1.21		?		
TiO_2	0.11	×	8.6	=	0.9	0.11	×	0.14	=	0.02
Al_2O_3	0.66	×	20.0	=	13.2	0.66	×	0.06	=	0.04
CaO	4.90	×	21.8	=	106.8	4.90	×	0.15	=	0.74
MgO	3.62	×	40.1	=	145.2	3.62	×	0.03	=	0.11
MnO	0.01		?			0.01		?		
CuO	0.84		?			0.84		?		
Na_2O	18.95	×	−2.4	=	−45.5	18.95	×	0.39	=	7.39
K_2O	2.06	×	4.6	=	9.5	2.06	×	0.33	=	0.68
PbO	0.07	×	9.9	=	0.7	0.07	×	0.08	=	0.01
SnO_2	0.08		?			0.08		?		
										11.6

Sum 1422.0°F

Fusion temperature = 1422°F (770°C)

Expansion coefficient = 6.4×10^{-6} in./in./°F (11.6×10^{-6} in./in./°C)

Percent expansion at 600°C = 0.67

TABLE 11–5

ESTIMATING PROPERTIES OF BLUE GLAZE WHICH WAS COMPOUNDED

Chemical Analysis			Fusion Coefficient of Oxide		Contribution to Fusion Temperature of Glaze (°F)	Chemical Analysis		Expansion Coefficient of Oxide		Contribution to Expansion Coefficient of Glaze
Na_2O	13.20%	×	−2.4	=	−31.7	13.20%	×	0.39	=	5.15
K_2O	1.93	×	4.6	=	8.9	1.93	×	0.33	=	0.64
CaO	6.73	×	21.8	=	146.7	6.73	×	0.15	=	1.01
ZnO	2.45	×	21.4	=	52.4	2.45	×	0.10	=	0.24
CuO	4.33		?			4.33		?		
B_2O_3	4.91	×	11.7	=	57.4	4.91	×	0.03	=	0.15
Al_2O_3	3.22	×	20.0		64.4	3.22	×	0.06	=	0.19
SiO_2	58.66	×	18.3	=	1073.5	58.66	×	0.04	=	2.35
			Sum		1372.0°F					9.73

Estimated Properties:

Fusion Temperature = 1372°F (744°C)
Expansion Coefficient = 5.41×10^{-6} in./in./°F (9.73×10^{-6} in./in./°C)

Measured Properties:
Glaze set point = 1058°F (570°C)
Fusion temperature = 1148°F (620°C)
Expansion coefficient = 4.74×10^{-6} in./in./°F (8.53×10^{-6} in./in./°C)
Percent expansion at 1058°F (570°C) = 0.465

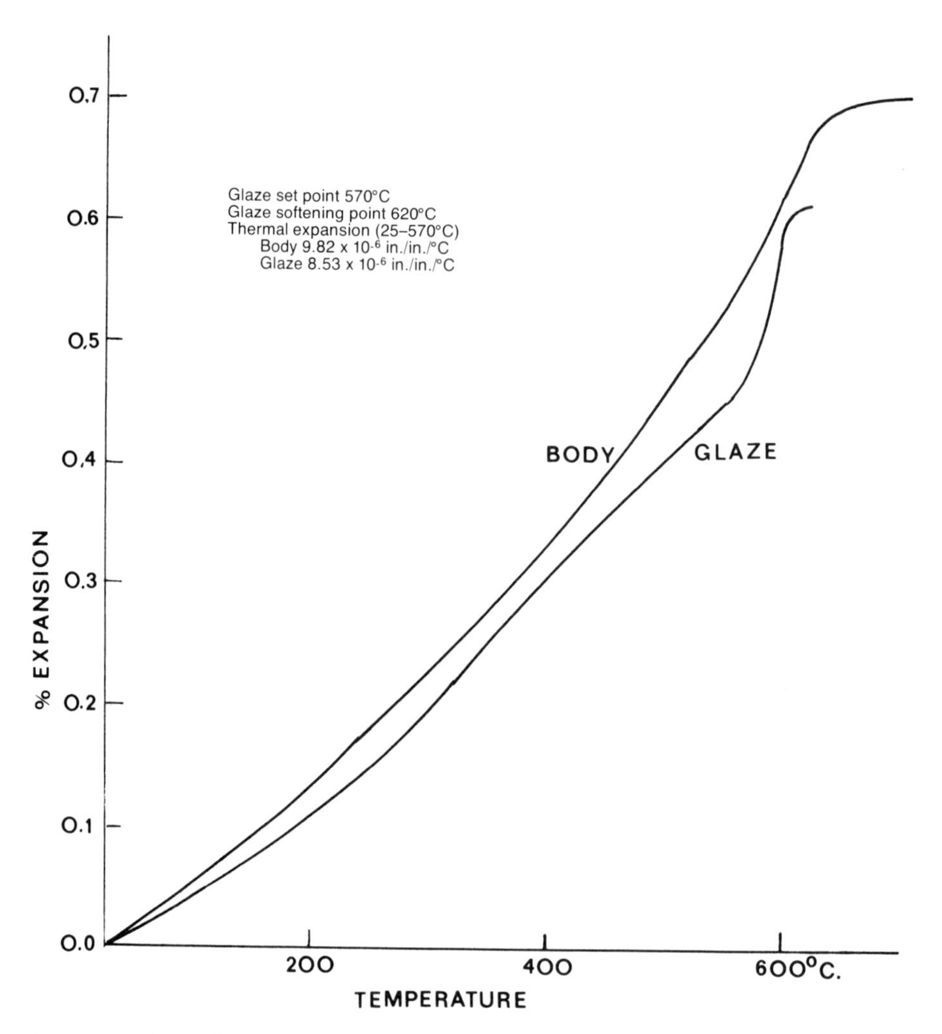

Fig. 11–6 Thermal expansion of Blue Hippo body and glaze (body fired to Cone 1).

dry handling strength and suspension of nonplastics in the casting slip. Twenty-five percent nepheline syenite was used for acceptable fired strength as well as for controlling the amount of expansion of the fired ware. Nepheline syenite glass has a high thermal expansion and also actively attacks the quartz, producing a siliceous glass with

TABLE 11–6

ESTIMATING THERMAL EXPANSION OF BLUE HIPPO BODY

Constituent in Unfired Body	Percent Unfired	Parts Fired	Percent Fired	Constituent in Fired Body	Percent	Expansion of Constituent at 600°C	Expansion Contribution
Clay	25.0	21.8	22.5	Mullite	15.0	0.265	.040
				Silica glass	7.5	0.031	.002
Potter's flint	50.0	50.0	51.6	Silica glass	25.8	0.031	.007
				Quartz	25.8	1.310	.338
Nepheline syenite	25.0	25.0	25.8	Nepheline syenite glass	25.8	0.520	.134
				Sum			.521

Estimated expansion of body at 600°C = .521%

Estimated expansion coefficient of body at 600°C = 9.08×10^{-6} in./in./°C (5.04×10^{-6} in./in./°F) quartz residue. Firing to a high temperature causes the quartz residue to be reduced giving a lower thermal expansion than if the ware is fired to a lower temperature.

The estimate of the thermal expansion of the body is shown in Table 11–6, assuming that half of the potter's flint changes to a siliceous glass and the other half remains as quartz. The body was fired over a range of temperatures to determine the proper firing temperature so that body and glaze would fit. Figure 11–6 shows the thermal expansion of the body fired at Cone 1 compared with the glaze.

The glaze has a set point at 570°C (1058°F) with a coefficient of expansion of 8.53×10^{-6} in./in./°C at this temperature. The body has an expansion coefficient of 9.82×10^{-6} in./in./°C at 570°C when the body is fired to Cone 1. Under these firing conditions the glaze has an expansion lower than the body by 13.1 percent, which is optimum for glaze fit. If crazing had begun to develop during production of the Blue Hippo, the ware could have been fired at a slightly lower temperature, and if shivering of the glaze had become apparent, the ware could have been fired slightly higher.

In the case of the development of the Blue Hippo body and glaze, the work in compounding many glazes and bodies was eliminated because of the capability of estimating the properties. As a result, early successful production of the Blue Hippo was achieved in the Met studios and available for public sale, thereby increasing interest in the Tutankhamen exhibit.

REFERENCES

1. Seger, H. A., *Collected Writings*, Chemical Publishing Co., Easton, Pa., 1902.
2. Rowland, D. H., "Porcelain for High Voltage Insulators," *Elect. Eng.*, 618–26 (June, 1936).
3. Croskey, C. D. and Rowland, D. H., Insulator, U.S. Pat. 2,287,976, June 30, 1942.

4. Winkelmann, A. and Schott, O., *Ann. Physik,* 51: 735 (1894).
5. Roberts, G. L., Salt, S., Roberts, W. and Franklin, C. E. L., "The Relationship between Chemical Composition and Physical Properties of Some Glazes in the System Na_2O-CaO-PbO-Al_2O_3-B_2O_3-SiO_2, *Trans. Brit. Ceram. Soc.,* 63: 553–602 (1964).
6. Norton, F. H., *Fine Ceramics, Technology and Applications,* pp. 191–93, McGraw-Hill Book Co., Inc., N.Y., 1970.
7. McLindon, J. D., "Estimation of Glaze Expansion from Chemical Compositions," B. S. Thesis, Alfred University, 1965.
8. West, R. and Gerow, J. V., "Estimation and Optimization of Glaze Properties," *Trans.* and *J. Brit. Ceram. Soc.,* 70, 7: 265–268 (1971).
9. Huff, D., "Electrical Porcelain Glaze-Body Interface," B.S. Thesis, Alfred University, 1964.
10. Merritt, C. W., "Defects in Artware Glazes," *J. Can. Ceram. Soc.,* 21: 42–44 (1952).
11. Marquis, J., "A General Paper on Glaze Defects," *J. Can. Ceram. Soc.,* 21: 45–50 (1952).
12. Turner, W. E. S., "Studies of Ancient Glass and Glass-making Processes," *J. Soc. Glass Tech.,* 436–56 (1953).
13. Binns, C. A., Klem, M. and Mott, H., "An Experiment in Egyptian Blue Glass," *J. Am. Ceram. Soc.,* 15: 271–72 (1932).
14. ———, *Ceramics and Man Through the Ages,* Aluminum Company of America, 1953.
15. Williamson, W. O., "The Scientific Challenge of Ancient Glazing Techniques," *Earth and Mineral Sci.,* 44, 3: 17–22 (1974).
16. Richie, M., "Models of Greatness," *Holiday Inn Companion,* 22–25 (January, 1977).

CHAPTER 12

SALT GLAZING

Salt glazing is a method of obtaining a glaze on clay products by throwing common salt into the kiln where it volatilizes and reacts with the surface of the ware to form a sodium-alumina-silicate glaze. The reactions involved are as follows:

1. $2NaCl + 2H_2O \rightarrow 2NaOH + 2HCl \uparrow$

The salt reacts with the water vapor present in the kiln atmosphere to form sodium hydroxide and hydrochloric acid fumes which are given off in the surrounding atmosphere.

2. $4NaOH + \text{Heat} \rightarrow 2Na_2O + 2H_2O \uparrow$

The sodium hydroxide formed in reaction 1 at the kiln temperature decomposes to form sodium oxide, Na_2O and water vapor which is evolved.

3. $2Na_2O + xAl_2O_3 \cdot xSiO_2 \rightarrow 2Na_2O \cdot xAl_2O_3 \cdot xSiO_2$

The sodium oxide reacts with the surface of the clay ware to form a sodium-alumina-silicate glaze.

Salt-glazed products, such as sewer pipe, silo blocks, vinegar jugs, butter crocks and coping tile, have been made for years and have proven very durable and resistant to chemical corrosion. Salt glazing is also an important technique used in the glazing and decoration of art pottery. Although salt glazing is no longer used extensively in commercial production, the technique is still of interest in the art pottery field.

Many variables affect the quality of a salt glaze. Defects such as dull, thin glazes, off-colored glazes, roughness, blistering, crazing and pig skinning may be evident. Recognition of these defects and the appropriate cures are of interest to the potter.

Salt-Glazing Practice

When red firing clays are used, the salt glazing is usually done at 1150 to 1200°C (2102 to 2192°F); when fire clay and stoneware clays are used, the customary salt-glazing temperature is 1260 to 1310°C (2300 to 2390°F). The ware is usually fire-flashed before salting. This is done by maintaining a reducing kiln atmosphere previous to salting. The reducing action of the kiln gases converts the ferric iron compounds to the ferrous state, in which condition the surface becomes more glassy, the body denser and less porous and a better salt glaze is produced on the surface.

Salt glazing may be done in coal-, oil- or gas-fired kilns. When coal is the fuel, the fires are allowed to burn clear, and the damper is partially closed before salting. Common rock salt is then thrown on each fire. The salting is repeated three to five times, and the kiln is allowed to cool slowly. Upon completion of the salting, the glaze appears gray because the iron compounds are in the black ferrous condition. The well-known mahogany brown color, characteristic of salt glazes, is developed by slowly cooling the ware in an oxidizing kiln atmosphere that converts the ferrous compounds into the reddish brown ferric forms.

The methods of firing and salting may vary considerably, depending somewhat on the clay used. If the clay contains soluble salts and has a tendency to produce rough glazes, it should be fired reducing above 540°C (1004°F). A reducing kiln atmosphere tends to decompose the calcium sulphate scum and, thus, facilitates the formation of better glazes. If the clay is high in calcium and/or magnesium, the ware should be fired as high as possible before salt glazing because such clays will only take dull, thin, discolored glazes at low temperatures. If it is necessary to use high-iron clays, they should be salt glazed at as low a temperature as possible because the body and glaze color would be too dark if fired at high temperatures. If the clay contains coarse iron particles and has a tendency to produce roughness, it should be fired oxidizing. This will oxidize any metallic iron particles to ferric oxide and minimize color variation in the surface.

The quality of a salt glaze is improved by adding water to the salt before introducing it into the fireboxes. Although water vapor is always present in the kiln gases from the combustion of the fuel, additional amounts are sometimes beneficial. Wood shavings are frequently added with the salt to introduce more water and also to increase the temperatures in the fireboxes. Oil or powdered coal may be mixed with the salt for the same reasons.

A vitrified body is most suitable for salt glazing, so it is advisable to fire and glaze the ware at or as near the vitrification temperature as possible. Adequate soaking to provide body maturity improves the quality of the salt glaze.

Body Compositions for Salt Glazing

Certain ceramic bodies will take a salt glaze, while others will not. This depends largely on the silica content and maturing temperature of the body. Barringer[1] found that bodies which take a good salt glaze usually have a composition corresponding to 1.0 mole of alumina to between 4.6 and 12.5 moles of silica. Mackler[2] found that the silica content may be as low as 3.3 and still take a good salt glaze. Kaolinite, $Al_2O_3 \cdot 2SiO_2 \cdot 2H_2O$, and talc, $3MgO \cdot 4SiO_2 \cdot 2H_2O$, are examples of compositions that could not be salt glazed because they do not have the proper alumina:silica ratio.

Clays containing 0 to 2.0 percent iron oxide usually take white to tan salt glazes.[3] Brown glazes are produced when the iron oxide content is 3.5 to 4.8 percent. Mahogany-colored glazes are produced when the iron oxide content is between 4.8 and 8.2 percent.

High-iron clays that also contain lime produce greenish yellow glazes. As low as 1.0 percent lime may prevent the clay from taking a bright glaze at low temperatures. Therefore, a high-lime clay can, best be salt glazed at high temperatures, but care must be exercised because of its short firing range.

Unlike lime, 1.5 percent magnesia improves the ability of the clay to take a salt glaze. However, when over 3.0 percent magnesia is present, it causes a dulling of the glaze. Clays containing over 3.0 percent magnesia tend to form crystalline salt glazes at 1260°C (2300°F).

The presence of 1.0 to 5.0 percent titanium oxide in the clay causes the salt glaze to be more brilliant.

If the clay is too low in silica, its ability to take a salt glaze may be materially improved by adding finely ground (—325 mesh) silica to the body. Selecting a clay lower in iron oxide will often improve the salt glaze. A 1.0 percent reduction in iron oxide is equivalent to the addition of 7.8 percent silica to a body which matures at 1150°C (2102°F) or 12.6 percent silica to a body maturing at 1210°C (2210°F) as far as the brightness of the glaze is concerned. Selecting a clay lower in alumina will often improve the glaze. A 1.0 percent reduction in alumina will often improve the glaze. A 1.0 percent reduction in alumina content is equivalent to a 2.7 percent increase in silica at all temperatures.

The colors of salt glazes obtained on clays containing certain percentages of iron oxide largely depend on the temperature at which the ware is fired and salt glazed. At 1145°C (2093°F), the iron oxide content should be 3.5 to 8.2 percent for a good dark brown color, while 1.5 to 3.8 percent iron oxide is ample to give the same color at 1285°C (2345°F). If the clay is too low in iron oxide to give a good color, this may be improved by firing under reducing conditions above 540°C (1004°F).

When using common salt, clays that overfire at 1125°C (2057°F) cannot be salt glazed because a higher temperature is required to cause the salt to combine with the body. It has been found, however, that when certain mixtures of borax and salt or boric acid and salt are used, good glazes are produced at temperatures as low as 1050°C (1922°F).[4]

If the clay or body contains more than 0.1 percent soluble calcium or magnesium salts, these salts may come to the surface and prevent the clay from taking a bright mahogany glaze; instead, they produce dull, thin, straw-colored glazes. To eliminate this scumming or migration of salts to the surface, it is common practice to add 0.1 to 0.3 percent of a barium compound, such as barium carbonate, barium hydroxide or barium chloride and soda ash. The barium reacts with the soluble calcium or magnesium salt rendering it insoluble and, therefore, incapable of migration to the surface. Much can be done to eliminate the troublesome effects of scumming by drying the ware as fast as possible.

The salt-glazing characteristics of a body are influenced somewhat by the method of body preparation. Although not of great importance in this day of air-floated clays, the fine grinding of raw clay is important. Also important is the removal of iron-bearing minerals and metallic iron particles. Such material causes rough glazes. Increasing the pugging time will often improve the glaze because this results in greater uniformity of the body and more even distribution of minerals that cause the roughness. Additions of 0.2 to 0.4 percent of soda ash or sodium chloride to the body will sometimes improve the quality of the salt glaze. These salts migrate to the surface during drying and aid in glaze formation by fusing with the surface.

Slip-Coated Salt-Glazed Ware

Finer grades of salt-glazed ware may be produced by applying engobes of varying compositions to the ware. Clays that are too impure or do not have the proper composition to salt glaze properly may be coated with a slip having the ability to take a salt glaze. These slips, or engobes, are composed of various mixtures of kaolin, flint, alkali silicates, glass frits, ball clay and metallic coloring oxides. They often contain 20 to 30 percent common salt and have a specific gravity of 1.3. They may be applied by dipping, spraying or brushing. A large variety of colors and textures can be produced by this process. The firing and glazing procedure must necessarily be altered to fit the properties of the slip used. An oxidizing atmosphere is usually necessary because reduction may cause the glaze to blister if metallic coloring oxides are present. Parts of the ware that are to be kept dull or unglazed are usually coated with slips high in alumina or magnesia.

Gray Salt Glazes

To get gray salt glazes it is essential to fire the ware high enough to produce a gray color in the body (1120–1150°C or 2048–2102°F). A transparent glaze is then applied over the body by means of four to six saltings. After salt glazing, the ware is cooled as rapidly as possible. An oxidizing kiln atmosphere with 150.0 percent excess air must be maintained throughout the salt-glazing period.

The clay must be finely ground to eliminate problems with iron specks that are especially noticeable against the gray background.

Colored Salt Glazes

As noted previously, colored salt glazes may be produced by applying a slip to the ware containing metallic-coloring oxides. Color may also be produced by mixing metallic-coloring chlorides or oxides along with the salt-glazing mixture. For brown glazes, approximately 2.0 percent manganese oxide or chloride is added to the salt. For blue on a buff burning body cobalt chloride is added. A large variety of colored glazes may be made in this manner. The color will depend on the amount and kind of coloring chloride and also on the color of the fired body.

Vapor Glazes Produced by Metals

The use of metallic zinc powder to produce green vapor glazed face brick is very similar to salt glazing ware with common salt. Zinc dust or flakes are introduced in the firebox at, or near, the end of the firing. This metal is volatilized and combines with the clay producing a yellow glaze over a blue body that results in a dark grayish blue green color. Although greens may be obtained on bodies with a wide range of compositions, which may have absorptions up to 8.4 percent, bodies having higher absorptions are not easily glazed. Bodies that take the best green glazes usually have absorptions below 2.0 percent, contain 8.0 to 18.0 percent iron oxide and also contain more than 6.0 percent alkalies. Because high alkalies improve the zinc green colors, it is good practice to initially salt glaze the surface with common salt before the zinc is introduced. Firing to high temperatures and cooling rapidly improves the likelihood of getting good green glazes.

When the lime content of the clay is above 3.0 percent, the color produced is usually a yellow green. Lowering the kiln temperature about 50°C (90°F) below the finishing temperature before zinc flashing produces better zinc greens than those developed by zinc flashing at the finishing temperature.

Desirable tans, brown, brick reds and grays are obtained by firing to a low temperature and cooling slowly.

Metallic bismuth also produces a glaze on ceramic bodies by the vapor glazing process although this metal is too expensive to be used except in special cases.

Improved Salt-Glazing Mixtures

It is possible to improve the surface quality of a salt glaze by the addition of other ingredients to the salt-glazing mixture. With some clays considerable improvement results from the addition of 2.0 percent zinc chloride to the salt. Lithium and potassium chlorides either alone or together with salt also result in some improvement. Probably the most practical method is the addition of borax or boric acid to the salt-glazing mixture.[4] Mixtures of boric oxide and sodium oxide fuse with the body at lower temperatures than when either of the fluxes is used alone. This explains why such mixtures make salt glazing possible at lower temperatures than when salt is used alone. Addition of 4.0 to 8.0 percent of borax or boric acid to the salt improves the thickness of the glaze, reduces roughness and pig skinning, darkens the color of the glaze, overcomes dull spots and reduces crazing.[4]

The best method of salting with boron compounds is to use salt alone for the first three saltings, then to use a mixture of salt and boron compounds and finally to use the boron compound alone for the final salting. The poorest method is to use boron compounds alone for the first salting and salt alone for the last three saltings.

Control of Salt Glazing

To obtain many of the effects described in this chapter, it is necessary to have control of the salt-glazing kiln. This is often a most difficult thing to achieve. Once a kiln has been used for salt glazing, its interior is loaded with an excess of salt, and each firing thereafter augments this situation. As a result, the glaze mixture used in any given firing may be overshadowed by what has gone before. Coloring oxides may not produce the expected results or may completely ruin the ware of a later firing. If a person is in the happy situation where he alone is in command of a salt glazing kiln, his experimentation will affect no one else. But when a number of people use the same kiln, each must be considerate of the others. In any case, duplication of results is most difficult under these circumstances. If experimentation with salt-glazing mixtures is to be at all extensive or effective, the building of a small salt-glazing kiln is recommended.

REFERENCES

1. Barringer, L. E., "The Relation Between the Constitution of a Clay and Its Ability to Take a Good Salt Glaze," *Trans. Am. Ceram. Soc.,* 4: 211–29 (1902).
2. Mackler, Tonind. Ztz., 251: 440 (1905).
3. Schurecht, H. G., "Clay Sewer Pipe Manufacture," *J. Am. Ceram. Soc.,* 6: 717–29 (1923); 7: 411–22, 539–47 (1924).
4. ——— and Wood, K. T., "The Use of Borax and Boric Acid Together with Salt in Salt Glazing," Bull. No. 2, Ceramic Experiment Station, SUNY College of Ceramics, Alfred, N.Y., 1942.

CHAPTER 13

TERRA SIGILLATA

History

Early Greek and Roman potters frequently employed terra sigillata as a very thin, opaque finish for pottery. It was usually red or buff, characteristic of the iron oxide colors, but sometimes brown or jet black. Although the pottery on which it was applied was usually porous, the coating was dense and impervious, with either a low gloss or a very high gloss similar to an enamel.

It is believed that the black coatings were made by a reducing fire. Frequently this color is found as open designs on a red body. Apparently, reduction changed the ferric oxide, Fe_2O_3, content of the coating to the black ferrous oxide, FeO, form. Because the coating was dense, it resisted oxidation during the cooling period, while the porous body was oxidized back to the red Fe_2O_3 color.

These coatings are dense, hard and resistant to chemical attack. Shards that have been buried for thousands of years show a gloss as good as the day the pieces were fired.

The making of terra sigillata coatings became a lost art. For many years the ceramic community attempted to duplicate these coatings, but, because of their extreme thinness and the fact that their composition was the same as the underlying body, analysis was difficult. For the first time, in the early 1930s Neumann[1] reported that the coatings were the same composition as the body. In 1936, Schumann[2] was granted a patent on terra sigillata coatings and described a method for making them. This involved separation of the fine particles from a clay (below 1μ) and their application to the body.

Preparation of Terra Sigillata

A proper terra sigillata coating depends on the presence of potassium-bearing clay minerals, either illite or muscovite.[3] The clay particles must be completely dispersed so that an overlapping fishscale structure is attained. During firing, the particles do not fuse but maintain their individual shapes. A terra sigillata coating for use at relatively low temperatures, such as 980 to 1100°C (approximately 1800 to 2000°F) requires the presence of fairly substantial amounts of illite. A high-temperature coating for use at 1260°C (2311°F) requires the presence of substantial amounts of finely divided muscovite. The low-temperature coatings with illite are easily found because most clays used for structural clay products contain substantial quantities of illite, but they fire to a red color. High-temperature coating clays are not as easily found, but some English china clays, such as HN or Bedminster, fire to a white color at 1260°C (2300°F).

Terra sigillata is prepared as follows:

Clay material	20.0	
Water	80.0	
Sodium metaphosphate	0.5%	(based on dry weight of clay)

The batch is ball-milled for four hours, transferred to large glass containers and allowed to settle for twenty hours; then the fine top fraction is siphoned off. This material may be used as is for application to the body, or it may be concentrated by drying. The amount of terra sigillata recovered usually amounts to 25 to 40 percent of the original weight of clay. This depends on the particle size distribution of the starting material. A ball clay will have a higher yield than a kaolin.

If other than natural colors are desired, the following batch may be used: dried terra sigillata (30%), coloring oxides (3%, based on dry weight of clay); water (70%). This batch is ball-milled for four hours. Coloring oxides that have been found to work satisfactorily with terra sigillata coatings include oxides of chromium, cobalt, copper, manganese as well as yellow and green body stains.[4] Most pleasing and unusual effects are produced.

Application To Body

For application to plastic or leather-hard bodies a specific gravity of 1.15 to 1.20 is satisfactory. Lower specific gravities are too wet to obtain satisfactory applications. For dry or bisque ware the material may require further dilution to a specific gravity of 1.05 to 1.10. Spraying works well for all body conditions. For dipping, it is well to moisten the body slightly by short immersion in water before dipping in terra sigillata. Otherwise pinholes may develop because of imperfect wetting and the entrapment of air bubbles. This same difficulty applies to bisque ware. The initial coating should be defect free because terra sigillata, unlike a glaze, will not flow or heal defects. It may also be applied by brushing. Such coatings are always applied very thinly by normal glaze standards. Due to their great opacity and hiding power 0.002 to 0.004 in. thickness is sufficient.

Any method of application results in a glossy coating. But especially high gloss, resembling a glaze, can be obtained by polishing the surface with a soft cloth while it is in the leather-hard condition. This gloss is retained during and after the firing process.[5]

Uses of Terra Sigillata Coatings

The characteristic of terra sigillata that makes it a dry coating, that is, it does not fuse at the firing temperature, presents an answer to some interesting problems. Blue colors used as underglaze have a washed-out appearance and also run to give poorly defined lines. The same color mixed either with a terra sigillata slip or the dry clay from the terra sigillata makes a vivid blue unattainable by other means, and the color does not run. Much less of the color should be used with terra sigillata as the clay particles act as an opacifier. A similar effect is obtained when using terra sigillata with pink stains. The colors become much more vivid, give a greater definition at the margin and do not run as much.

The use of terra sigillata as an interfacial coating between the body and glaze also offers numerous advantages. The cause for most of the bubbles in a glaze is reaction of quartz with the glaze. Inasmuch as the terra sigillata contains no quartz and does not readily react with the glaze, there are fewer bubbles in a glaze placed on the terra sigillata coating than there are in one placed directly on the body.

Low-expansion interfacial coatings have been used for a number of years to strengthen electrical porcelain, but most of the coatings contain large amounts of talc in order to attain the low thermal expansion desired.[6–9] A body with an expansion coefficient of 5.6×10^{-6} in./in./°C should have a glaze with an expansion coefficient of 4.75 to 5.0×10^{-6} in./in./°C, and the expansion coefficient of the coating should be less than 4.0×10^{-6} in./in./°C. However, the results are quite selective so that inexplicably some coatings only work properly with some body-glaze combinations. A terra sigillata coating has an expansion coefficient of about 3×10^{-6} in./in./°C and appears to work generally with many body-glaze combinations. The increase in strength of hotel china cups was tripled and the modulus of rupture of glazed electrical porcelain rods was increased by 30 percent by employing an interfacial terra sigillata coating.

Such coatings will not flow, drip or stick to each other or to the kiln furniture. It is not necessary to leave a dry foot on a piece or to separate the ware from the setter slab with a layer of sand. The ware may be fired in the open, unprotected by saggars. However, it should be pointed out that these coatings are subject to the same oxidation-reduction color variations as any clay body. Consistent results require the usual attention to temperature-time-atmosphere control.

REFERENCES

1. Neumann, B., "Glaze of Terra Sigillata," *Sprechsaal*, 65, 14: 253–55; 15: 273–75; 16: 291–93 (1932); *Ceram. Abs.*, 11, 8: 461 (1932).
2. Schumann, T., (Schutte A. G. fur Tonindustrie) German Patent No. 626,112, February, 1936; *Ceram. Abs.*, 15, 8: 249 (1936).
3. Bestwick, J. D. and Smith, T. A., "The Surface Finish of Samian Ware," *Sci. and Archaeol.*, 12: 21–31 (1974).
4. Coffin, L. B., "Colored Terra Sigillata Coatings for Building Materials," *Bull. Am. Ceram. Soc.*, 37: 10 (1958).
5. Amberg, C. R., "Terra Sigillata, Forgotten Finish," *Ceram. Ind.*, 51: 6 (1948).
6. Rowland, D. H., "Porcelain for High Voltage Insulators," *Elect. Eng.*, 618–26 (June, 1936).
7. Smothers, W. J., Glazed Insulators Which Comprises a Body and Primary Coat under the Glaze Which has a Lower Coefficient of Thermal Expansion than the Ceramic Body, U. S. Patent 3,024,303, March 6, 1962.
8. Bestwick, J. D. and Smith, T. A., "The Surface Finish of Samian Ware," *Sci. and Archaeol.*, 12: 21–31 (1974).
9. Huff, D., "Electrical Porcelain Glaze-Body Interface," B.S. Thesis, Alfred University, 1964.

CHAPTER 14

THERMAL SHOCK THEORY

Ceramic bodies capable of withstanding sudden changes in temperature are of interest to the potter, particularly for ware designed for oven or top-of-the-stove use. But before we discuss thermal shock resistant bodies, let us examine the many factors involved in the thermal shock behavior of ceramic materials.

Ceramic materials break in thermal shock because tensile stresses develop which exceed the strength of the body. These stresses are caused by differences in the thermal expansion which, in turn, are caused by a temperature gradient through the wall of the piece.

Figure 14–1 A shows a wall section with one face hot and the opposite face cold. Because the hot face expands more than the cold, tensile stresses will result on the cold, and when these stresses exceed the tensile strength of the body, a crack will develop on the colder side.

Several physical properties of a material are involved in the prediction of the temperature difference that a piece can withstand without failing. These have been interrelated by the following expression:

$$\Delta T_f = \frac{KS_t(1 - \mu)}{\alpha E}$$

where:

ΔT_f = The temperature difference causing failure,
K = Thermal conductivity of the material,
S_t = Tensile strength of the material,
μ = Poisson's ratio,
α = Thermal expansion,
E = Modulus of elasticity.

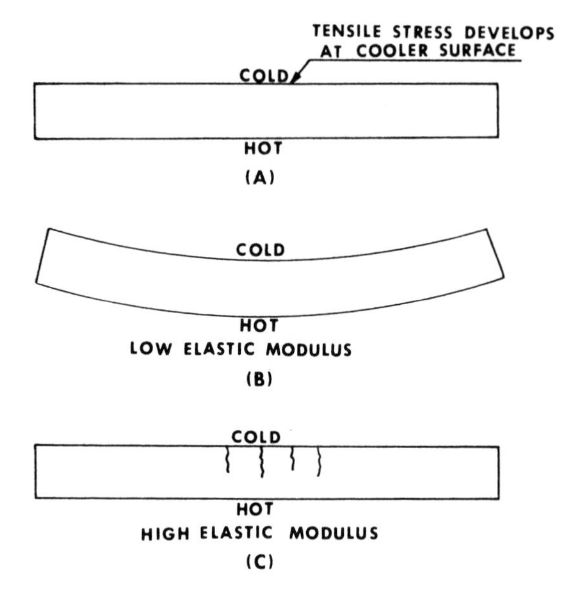

Fig. 14–1 A ceramic slab heated on one side. (A) Development of tensile stress on cold surface, (B) bending of a material having a low modulus of elasticity, (C) crack development at cold surface in brittle ceramic materials.

Kingery[1] and Crandall and Ging[2] have made refinements in the above expression to cover special cases and conditions, but for our purpose the above expression is sufficient. Let us examine these physical properties individually to ascertain their role in thermal shock behavior.

Strength (S_t)

Although tensile strength is an important factor in determining the amount of tensile stress a body can withstand before a crack develops, there is not much one can do about improving strength. Of course, proper body composition and proper maturing of the body develops the best strength, but, in most cases, the strength of ceramic bodies falls within a rather narrow range. None of the strengths developed is capable of withstanding the tremendous stresses resulting from thermal expansion; for example, the pressure required to prevent an increase in volume when MgO is heated from room temperature to 100°C (212°F) is 70,000 lbs./in.^{2} Ceramic bodies do not have the strength to cope with this amount of stress. Consequently, one simply recognizes that tensile strength is a factor, however small, and "lives with" the strength attainable in particular body compositions. The strengths of some common ceramic materials are given in Table 14–1.

Poisson's Ratio (μ)

This property relates to the elastic properties of the body. It is defined as the ratio between the decrease in width and the increase in length of a cross section of a bar as it undergoes tensile stress. Ceramic materials at room temperature flow very little; therefore, Poisson's ratio is small. For most clay bodies it has a value of 0.30. For glass it is slightly lower and for silica glass[3] it has a value of 0.15. This also is a property that cannot be controled.

TABLE 14–1
TENSILE STRENGTH OF SOME CERAMIC MATERIALS

	Tensile Strength lbs./in.^{2}
Electrical porcelain	13,000
Alumina	30,000
Steatite	11,000
Glass	7–13,000
Quartz glass	12,000
Stoneware glaze	2,000
Dried clay	1,400

Modulus of Elasticity (*E*)

This property is a measure of the capacity of a body to withstand stress without deformation. Ceramic bodies, being brittle, deform very little and can withstand high stress without deformation. Therefore, they have a high modulus of elasticity. A material, such as rubber, which deforms easily with small applied loads has a very low modulus of elasticity. If a material has a low modulus of elasticity, it will bend when subjected to stress developed by temperature difference. The cooler surface would become concave, thus relieving the stress (Figure 14–1 B). This is the situation one observes in bimetallic springs used as temperature sensors in thermostats. Ceramic materials having high moduli of elasticity do not bend to relieve the stress but simply crack (Figure 14–1 C). This property also is not under our control. Table 14–2 gives the modulus of elasticity values for some common ceramic materials.[3]

Thermal Conductivity (*K*)

If there were no temperature difference through the wall of a piece, there would be no differential expansion and no stresses would be

TABLE 14–2
MODULUS OF ELASTICITY OF SOME CERAMIC BODIES[3]

Body	$E \times 10^6$	Body	$E \times 10^6$
Fire clay	0.5–1.0	Leadless glaze	8
Brick	0.7–1.2	Lead glaze	12
Earthenware body	5.0	Glass	10–25
		Porcelain	8–12
Tile body	2.0		

developed. Any temperature gradient that is developed depends on the thermal conductivity of the material. This property is a measure of the ability of the material to transfer heat through a given wall thickness. Metals have high conductivities as compared with most ceramic materials. This results in lower temperature differences being generated through comparable wall thicknesses, and accounts for metals having better heat shock resistance than ceramic materials. Figure 14–2 gives the thermal conductivity of some ceramic materials.[4]

As is apparent, the thermal conductivity of the types of bodies of interest to the potter are at the low end of the thermal conductivity scale in the 1 to 4 range. These values drop even further as bodies are made porous. On the other hand, additions of oxides, such as alumina and magnesia, may increase the thermal conductivity of a body slightly, but the overall effect of modest additions does not result in appreciable improvement. Beryllium oxide additions are ruled out because of the cost of the material and its extreme toxicity. Thermal conductivity then, along with all of the previously discussed properties, is for all practical purposes not subject to manipulation with the tools and methods available to the potter.

Thermal Expansion (α)

Thermal expansion, the last remaining variable to be discussed, is the one property that can be varied and controlled by the potter. This is a measure of the change in length of a material as it is heated from one temperature to another. It is usually expressed by the expansion coefficient in inches per inch per degree Celsius. A material such as alumina has an expansion coefficient of 8×10^{-6} in./in./°C. This means that if a 1-in. rod of alumina is heated 1°C, its length will then be 1.000008 in. If it were heated to 1000°C, its length will be 1.008 in.

Referring to Figure 14–1, it is apparent that if a material had no expansion or contraction when heated, it would not matter what the

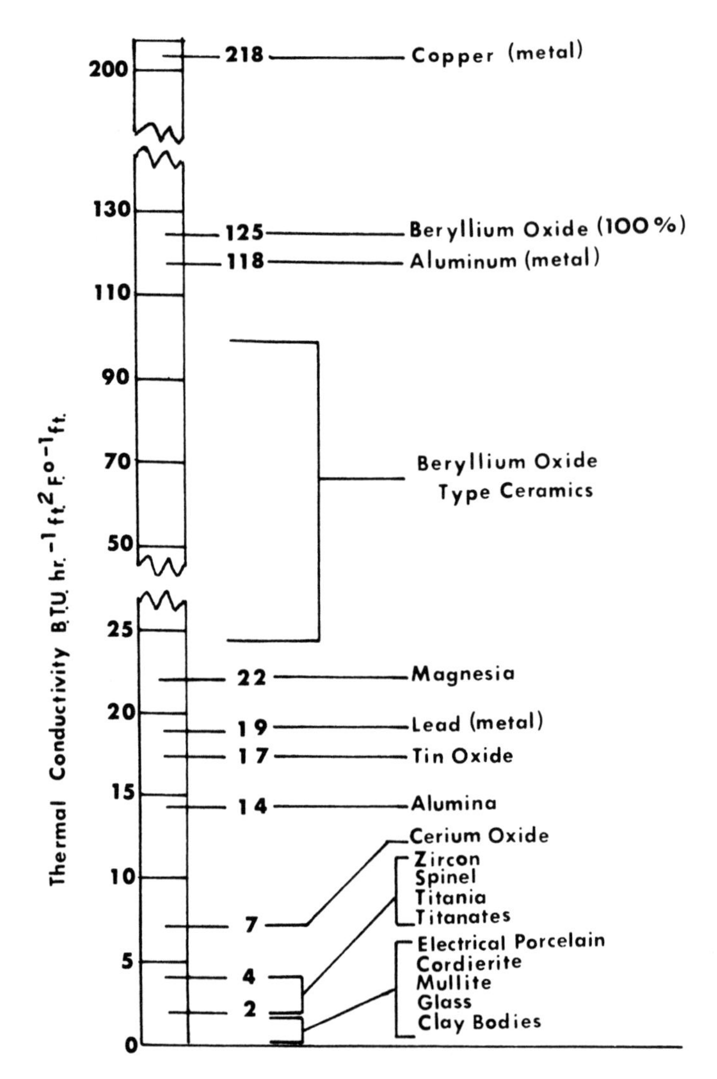

Fig. 14–2 Thermal conductivity for some ceramic materials.

temperature difference was between the hot and cold face because no stress would be produced. Silica glass having the very low expansion coefficient of 0.5×10^{-6} is almost completely insensitive to temperature changes.

Fortunately, ceramic materials have a range of expansions depending on their mineralogical and chemical composition. Figure 14–3 shows some of the expansion coefficients for ceramic bodies.[4] Some of the more interesting are plotted on a greatly enlarged scale in Figure 14–4.

Of particular interest to the potter are compositions that have low expansions in the range 0 to 3 $\times$ 10^{-6}. Bodies having such expansion characteristics are truly thermal shock resistant. It is apparent that such bodies must be based on the crystalline phases of cordierite, $2MgO{\cdot}2Al_2O_3{\cdot}5SiO_2$, silica glass, spodumene, $Li_2O{\cdot}Al_2O_3{\cdot}4SiO_2$, or petalite, $Li_2O{\cdot}Al_2O_3{\cdot}8SiO_2$. Chapter 15 discusses the development of such bodies.

Small changes in thermal expansion can be obtained by making additions of a lower expanding material to the body. A typical example is the replacement of silica by alumina. In this case, the crystalline silica, having a high expansion, is replaced by the lower expanding alumina. The resulting body will have a lower expansion than the original. This type of replacement is strictly a physical one,

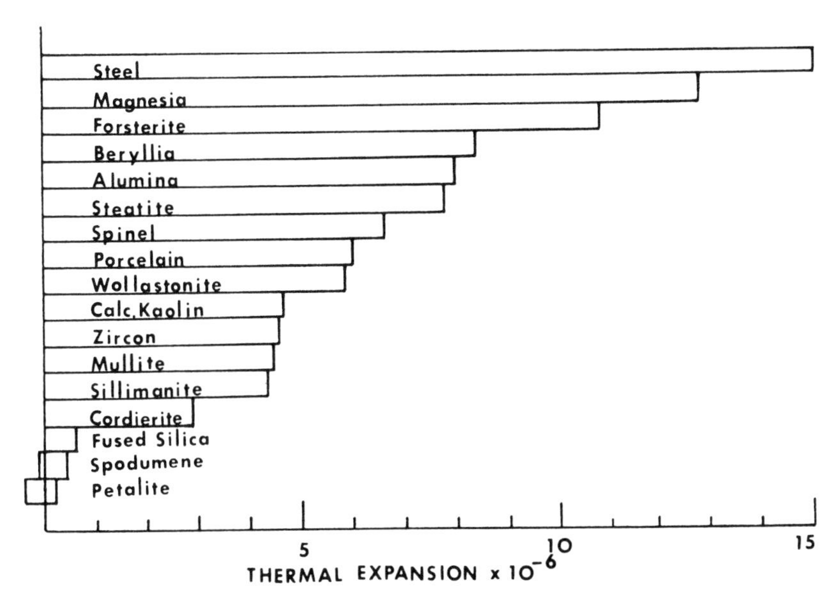

Fig. 14–3 Thermal expansion for some ceramic materials.

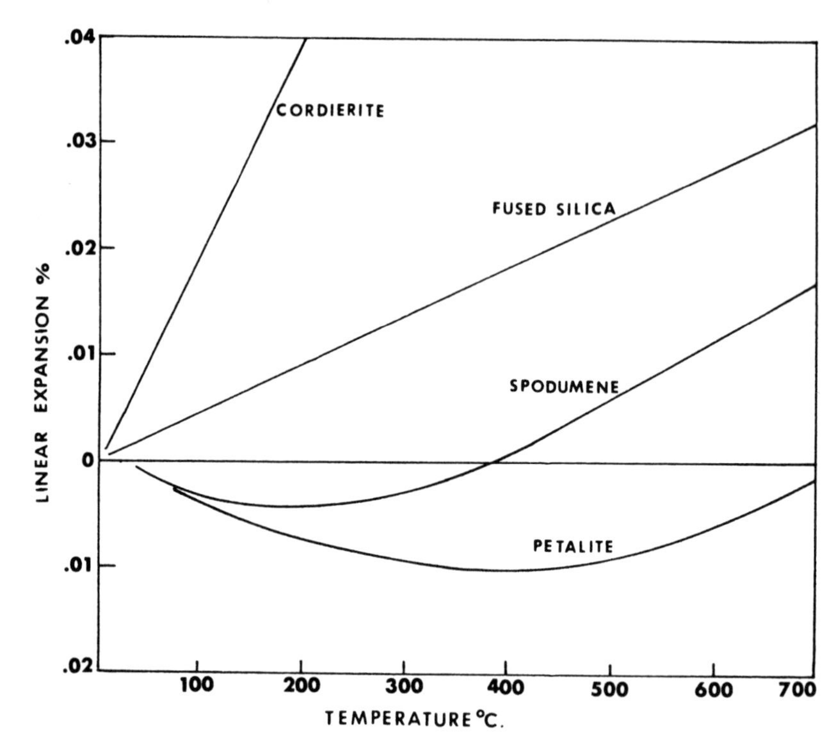

Fig. 14–4 Thermal expansion of some low-expanding materials showing unusual expansion for the lithium-alumina-silica minerals.

and the resulting expansion will be intermediate between the two components involved.

If the addition reacts with other body components to form a new phase having different expansion characteristics, a considerable change may result. An example of this type of addition is talc. It is well known that the addition of talc to clay bodies decreases the expansion and results in a body that is much more resistant to thermal shock. The talc, $Mg_3Si_4O_{10}(OH)_2$, reacts with the clay constituent of the body to form cordierite, $Mg_2Al_3(AlSi_5)O_{18}$, which has a low expansion. This reaction occurs at a temperature of approximately 1350 to 1450°C (2462 to 2642°F).

Other Factors Influencing Thermal Shock Resistance

CRACK PROPAGATION

The ability of a body to accommodate the thermal stresses built up is difficult to measure. Once the stress is sufficient to start a crack, the propagation and enlargement of that crack will depend on the physical characteristics of the body. A glassy body is subject to crack propagation much more than a crystalline body. In the latter case the path of the crack in the glassy phase is interrupted occasionally by crystals. Of much greater influence, however, is the effect of porosity. In a porous body the crack path is interrupted by an air space. This is illustrated in Figure 14–5.

It has been observed that porous insulating fire brick will stand greater thermal shock than the dense variety. They withstand a greater number of heating and cooling cycles before obvious spalling occurs. This does not mean that they do not crack. Cracks have been shown to develop, but the cracks are not continuous and do not immediately propagate through the piece, which then results in the loss of a portion of the brick. The crack starts at the surface and propagates until it reaches a pore. The pore interrupts the crack and probably relieves some of the stress developed. On the next heating and cooling cycle, the crack may continue further through the body and eventually the

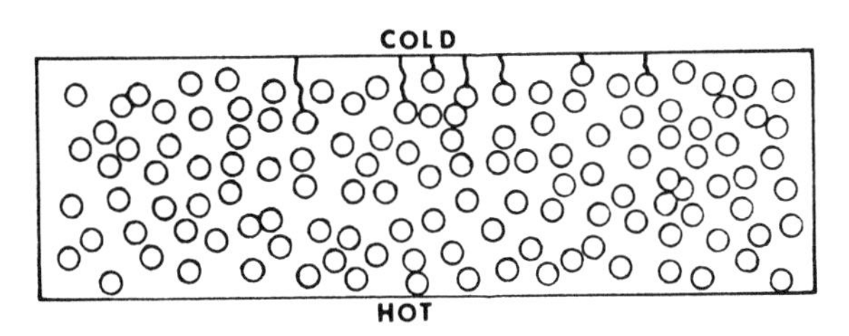

Fig. 14–5 Crack propagation in a porous body showing interruption of crack by a pore. Further growth of the crack would be expected on next heating and cooling cycle.

body will fail. This interpretation is supported by the fact that porous materials gradually lose their strength with repeated heating and cooling.

CRITICAL WALL THICKNESS

For a given thermal conductivity, it is apparent that the temperature difference through a wall will depend on the thickness of the wall. The thinner the wall the smaller the temperature difference and, therefore, the smaller the tensile stress on the cold surface.

It follows that the thinner the wall the greater the thermal shock resistance of the piece. There will be a certain critical wall thickness below which the piece cannot be broken in thermal shock. Even though such thin wall thicknesses may not be attainable in pottery manufacture, it is an important concept and much can be done to improve the thermal shock properties of a ceramic material by making thin cross sections. In reducing wall thickness, other factors, such as strength, forming, handling and firing problems, must be considered. But with proper control of the expansion properties, wall thickness becomes of secondary importance.

REFERENCES

1. Kingery, W. D., "Thermal Stress Resistance of Ceramic Materials," *J. Am. Ceram. Soc.*, 38: 3 (1955).
2. Crandall, W. B. and Ging, J., "Thermal Shock Analysis of Spherical Shapes," *J. Am. Ceram. Soc.*, 38: 1 (1955).
3. Searle, A. B. and Grimshaw, R. W., *The Chemistry and Physics of Clays*, 3rd ed., Interscience Publishers, Inc., N.Y., 1959.
4. Smoke, E. J. and Koenig, J. H., "Thermal Properties of Ceramics," *Eng. Res. Bull.* No. 40, College of Engineering, Rutgers University, New Brunswick, N.J., 1958.

CHAPTER 15

THERMAL SHOCK BODIES

Ceramic ware made to withstand the type of thermal shock encountered in the home falls into two categories: ovenware and flameware.

Norton[1] lists the following requirements for ovenware:

1. Must withstand quenching from 150°C (302°F) into cold water.
2. Should have smooth, hard, noncrazing surface.
3. Must be capable of durable decoration.
4. Must have reasonable mechanical strength.

Most suitable ovenware bodies are based on the development of cordierite, which has a low expansion. Stoneware and clay-grog bodies are borderline as far as thermal shock resistance is concerned. For ovenware, the thermal expansion coefficient should be less than $3 \times 10^{-6}/°C$.

Ovenware will resist thermal shock breakage provided the constituents in the fired ware do not have discontinuities in their thermal expansion curve. The most notable constituents of stoneware and clay-grog bodies having such discontinuities are quartz and cristobalite. The quartz discontinuity is above oven temperatures, so cristobalite is probably the constituent that is most responsible for thermal cracking in ovenware. Hotel china ware may crack at high dishwasher temperatures because it contains approximately 1 percent cristobalite.

Raw materials containing only a bare minimum of quartz should be used. Materials that introduce large amounts of quartz into the body, such as some feldspars and ball clays, should be avoided. Constituents that contain lime or iron have a tendency to change quartz to cristobalite during firing, so these constituents should be avoided also. If possible, nepheline syenite should be used instead of feldspar to dissolve small amounts of quartz which may have been introduced inadvertently into the body composition.

A body composition containing no quartz produces fired ware with a thermal expansion so low the ware is difficult or impossible to glaze. To produce ware having an expansion sufficiently high to match a glaze, a minimum of quartz is sometimes used in the body (10% or less), and the constituents that contain lime or iron are reduced as low as possible so that little or no quartz is converted to cristobalite. Extremely fine quartz grinds should be avoided, as the fine-grained quartz might also convert to cristobalite during firing. Body constituents that have a high linear expansion, such as calcined alumina and nepheline syenite, may be used to reduce the possibility of thermal shock and yet produce ware that may have a sufficiently high thermal expansion to be glazed.

Figure 15–1 shows the various types of low-expansion bodies and the recommended thermal expansion areas for ovenware and flameware. Flameware must stand direct contact with flame temperatures and electric heating elements and, therefore, must be more resistant to thermal shock than ovenware. Such ware should have expansions below $2 \times 10^{-6}/°C$.

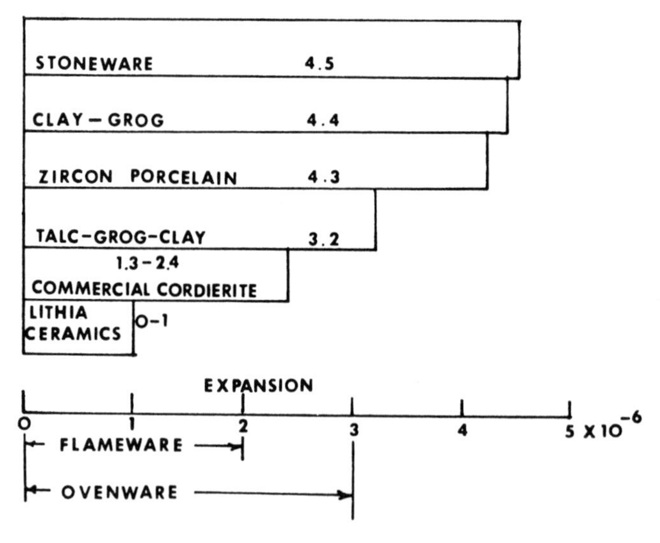

Fig. 15–1 Thermal expansions for some of the more shock resistant ceramic bodies with the desirable thermal expansion range for ovenware and flameware.

Because the only bodies well within the expansion limits designated are cordierite and lithia ceramic compositions, the remainder of this chapter will deal with these bodies and low-expansion glazes for these bodies.

Cordierite Bodies

Cordierite, $2MgO{\cdot}2Al_2O_3{\cdot}5SiO_2$, is a crystalline material having a low thermal expansion. It is formed when minerals that contain MgO, such as talc, are fired together with clay. The resulting formation of cordierite lowers the expansion and improves the thermal shock resistance of talc-clay-grog bodies. Pure cordierite bodies have been difficult to produce because the cordierite composition has a very short firing range, as little as 5°C (9°F). This explains why the potter has had little interest in cordierite bodies.

Recent work has been reported on cordierite ceramic bodies having improved or longer firing ranges. Jelacic[2] has studied the effect of various additions on the extension of the firing range and on expansion and thermal shock resistance. The basic dense cordierite composition consisting of MgO 22.6 percent, kaolin 69.4 percent and silica 8.0 percent has an abrupt melting point and considerable shrinkage, making it extremely sensitive and difficult to fire properly. The effect of lead silicate, lead borate, barium carbonate and potash feldspar additions was studied, as well as additions of forsterite and cordierite grog. The compositions of the starting batches, the additions and the resulting properties are given in Table 15–1.

The additions of lead silicate extended the vitrification range from 1370 to 1410°C (2498 to 2570°F). Samples with sharp corners appeared satisfactory and retained their sharpness throughout the firing range. Lead borate additions resulted in a vitrification range of 1280 to 1320°C (2336 to 2408°F), but above this bloating occurred. Barium carbonate additions were not considered satisfactory.

The most interesting of all the additions tested was the potash feldspar. A 10 percent addition resulted in a vitrification range of

TABLE 15–1
CORDIERITE BODY COMPOSITIONS[2]

	1	2	3	4	5
Magnesite	22.6	–	22.6	–	–
Kaolin	69.4	43.0	69.4	43.0	43.0
Silica	8.0	–	8.0	–	–
Cordierite-grog (porous)	–	45.4	–	45.4	–
Cordierite-grog (dense)	–	–	–	–	45.4
Forsterite	–	11.6	–	11.6	11.6
	100.0	100.0	100.0	100.0	100.0
Additions					
$PbSiO_3$	10.0	6.0	–	–	–
$BaCO_3$	–	6.0	–	–	–
Potash spar	–	–	10.0	12.0	12.0
Maturing temperature, °C	1360.0	1340.0	1340.0	1360.0	1360.0
Shrinkage, %	4.0	10.0	9.0	3.0	4.0
Absorption, %	0.7	0.0	0.6	0.2	0.5
Expansion $\times 10^{-6}$/°C	1.58	2.42	2.04	1.90	1.39

1300 to 1360°C (2372 to 2480°F), and the bodies were found to be self-glazing. Above 1360°C, (2480°F), bloating occurred.

Because of the interesting properties developed by the feldspar additions, compositions were made up varying from porcelain to cordierite. These bodies and the resulting properties are shown in Table 15–2.

With continually increasing additions of MgO to the clay-feldspar body (1), it is noted that the coefficient of expansion decreases continuously from 4.28 to 1.58 $\times$ 10^{-6}/°C for the 100 percent cordierite (6). Jelacic[2] suggests that body No. 7 is the most promising.

TABLE 15–2
PORCELAIN-CORDIERITE BODIES AND PROPERTIES[2]

	1	2	3	4	5	6	7
Kaolin (Premier)	71.0	71.0	70.0	69.0	68.0	68.0	63.1
Feldspar	27.0	23.0	17.0	12.0	6.0	0.0	9.1
Clay (Nerof)	2.0	2.0	2.0	2.0	2.0	2.0	0.0
Silica	0.0	0.0	2.0	4.0	6.0	8.0	7.3
Magnesite	0.0	4.0	9.0	13.0	18.0	22.0	20.5
Fusion point, °C	1600.0	1560.0	1540.0	1430.0	1410.0	1390.0	1380.0
Absorption, %	0.4	0.5	0.6	1.2	1.3	19.5	0.8
Mineral composition	M	M	M	C	C	C	C
M-Mullite, Q-Quartz	Q	Q	C	M	M	G	G
C-Cordierite, G-Glass	G	C	Q	Q	G	Q	Q
		G	G	G	Q		
Expansion × 10^{-6}/°C	4.28	3.96	3.21	2.37	1.90	1.58	2.20

Several different shapes were made from composition No. 7 and fired at 1370°C (2498°F). There was no deformation, the fired color was white to bluish white and a brilliant, glossy self-glaze with no blemishes developed. Casseroles containing water were boiled to dryness and placed directly on electric heating coils without failure.

Smoke and Koenig[3] report on a similar method of improving the firing range of a cordierite body. This involves the preparation of a calcine from a portion of the batch, firing this just short of densification, grinding and adding it to the remainder of the batch. This results in a 50°C (90°F) firing range as compared to 5°C (9°F).

CALCINE

Oxide		Batch	
MgO	36.0%	Talc	17.2%
Al_2O_3	18.0	Kaolin	33.2
SiO_2	35.9	$MgCO_3$	42.5
$B_2O_3 \cdot P_2O_5$	10.1	$B_2O_3 \cdot P_2O_5$	7.1
	100.0		100.0

Calcine above batch at 1000°C

The body composition is as follows:

Oxide		Batch	
MgO	10.6%	Calcine (above)	34.2%
Al_2O_3	33.9	Kaolin	56.3
SiO_2	50.2	Ball clay	9.5
$B_2O_3 \cdot P_2O_5$	5.3		100.0
	100.0		

This body matures between 1300 and 1350°C (2372 and 2462°F). When fired at 1325°C (2417°F), it has an expansion of 1.95×10^{-6}/°C and a modulus of rupture of 13,300 psi.

The need to make the cordierite-grog can be eliminated by using calcined clay-grog, as in the following formula reported by Smoke and Koenig.[3]

Talc	20.4%
Kaolin	36.7
Calc. kaolin	36.7
$BaCO_3$	6.2
	100.0

This body matures at 1350°C (2462°F) and has an expansion of 1.95×10^{-6}/°C.

If an excess of $BaCO_3$ is added, a lower thermal expansion results. Smoke and Koenig[3] attribute this to the development of a "cordierite-like" crystalline phase having the composition $BaO \cdot 2MgO \cdot 4.7Al_2O_3 \cdot 12.3SiO_2$.

Talc	15.7%
Kaolin	36.0
Calc. kaolin (1000°C)	36.0
$BaCO_3$	12.3

Fired at 1330°C (2426°F), this body has an expansion of 1.65×10^{-6}/°C.

Thiess[4] reports several dense cordierite body compositions fired in the 1350 to 1390°C (2462 to 2534°F) range. Three compositions follow:

	1	2	3
Talc	36.0	36.0	36.0
Ball clay	29.0	29.0	29.0
Alumina	19.0	19.0	19.0
ZnO	–	3.2	–
Feldspar	16.0	12.8	–
Nepheline syenite	–	–	16.0

These are the properties after firing at Cone 13, 1350°C.

Transverse strength, psi.	9950	10,020	9137
Color	Gray	Gray	Gray
Absorption, %	0	0	0
Expansion $\times 10^{-6}$/°C	2.96	2.78	2.61

Although the preceding studies have done much to improve cordierite bodies, these compositions still hold little interest for the potter because of their high-firing temperatures, short-firing ranges and overall fabrication problems. For these reasons, low-expansion lithia ceramic bodies provide the basis for most of the thermal shock ware made by the potter.

Lithia Ceramics

The lithium alumina silicate minerals, spodumene ($Li_2O{\cdot}Al_2O_3{\cdot}4SiO_2$), petalite ($Li_2O{\cdot}Al_2O_3{\cdot}8SiO_2$) and eucryptite ($Li_2O{\cdot}Al_2O_3{\cdot}2SiO_2$), exhibit abnormally small expansion coefficients. Eucryptite has a negative expansion of -9×10^{-6}/°C.[5] Negative expansion of this magnitude is as bad, as far as thermal shock resistance is concerned, as positive expansion. Therefore, the following discussion will concentrate on the petalite-clay and spodumene-clay bodies, both of which are especially adaptable and easy to use. Expansion coefficients can be made to order from 0 to 2×10^{-6}/°C with little difficulty.

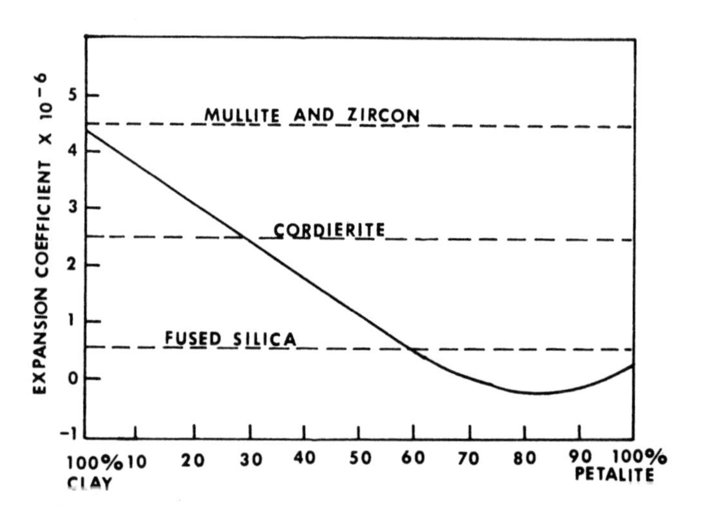

Fig. 15–2 Thermal expansion versus composition for clay-petalite bodies.[6]

The expansion behavior of simple clay-petalite bodies has been reported. The thermal expansion as a function of composition is shown in Figure 15–2. Such bodies have a wide useful range of low-expansion compositions. Table 15–3 shows the properties of some clay-petalite bodies.

These data indicate the composition range of 35 to 65 percent clay would be the best to work with.

Fishwick, Van Der Beck and Talley[7] describe several petalite-clay and spodumene-clay compositions. The effect of composition on expansion properties is shown in Figure 15–3. Table 15–4 gives the properties of several compositions fired to 1350°C (2462°F).

A change in firing temperature has a considerable effect on the resulting thermal expansion properties of the body. Table 15–4 gives properties for samples fired at 1350°C (2462°F) only. The reader is referred to Fishwick et al.[7] for further detailed information. In the 1350 to 1410°C (2462 to 2570°F) range, remarkably low and even negative expansion coefficients were observed for bodies containing 50 to 60 percent spodumene or 70 percent petalite.

To duplicate reported expansion properties of lithia ceramic bod-

TABLE 15–3
PETALITE-CLAY BODY COMPOSITIONS[6]

	1	2	3	4
Petalite (200 mesh)	35%	55%	65%	90%
China clay	–	45	35	10
Plastic fire clay	65	–	–	–
Properties when				
fired at cone	6.0	13.0	13.0	11.0
Absorption, %	18.5	2.8	9.8	27.9
M. of R., psi	1575.0	5900.0	2775.0	2750.0
Exp. $\times 10^{-6}$/°C	1.49	0.78	0.16	−0.01

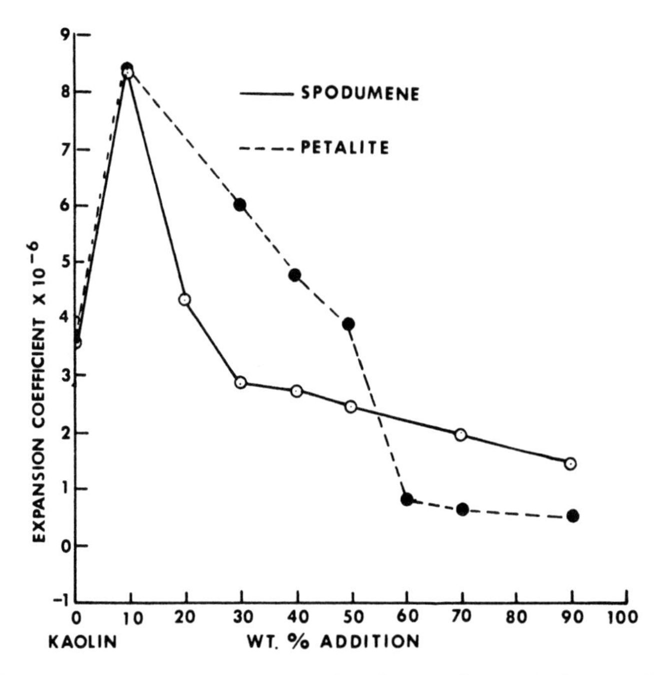

Fig. 15–3 Expansion versus composition for clay-petalite and clay-spodumene bodies.[7]

ies, one must exactly duplicate firing temperatures and times. This is dramatically illustrated not only by the varying expansion coefficients in the previously mentioned Fishwick[7] paper, but also by the findings of Maki and Tashiro.[8] They fired samples of petalite 60 percent and kaolin 40 percent to two different temperatures, 1290 and 1310°C (2354 and 2390°F), and soaked them at these temperatures for 30 minutes. The linear expansion of the body fired at 1290° C was 1.1×10^{-6}/°C, but for the body fired at 1310°C, only 20°C higher for the same length of time, the coefficient of expansion was 3.5×10^{-6}/°C. The first fired body would be suitable for a top-of-the-stove application, but the second would be completely unsatisfactory. This point is illustrated in Figure 15–4 and is a most important consideration when dealing with lithia ceramic bodies.

TABLE 15–4
PROPERTIES OF SOME PETALITE-KAOLIN AND SPODUMENE-KAOLIN COMPOSITIONS[7]

	Composition		Expansion × 10^{-6}/°C	Firing Shrinkage	Modulus of Rupture, psi
	Spodumene	**Kaolin**			
1		100	3.8	11.3	2960
2	10	90	7.8	8.9	6700
3	20	80	4.4	10.5	7220
4	30	70	2.4	9.0	6720
5	40	60	1.6	7.1	6540
6	50	50	1.9	6.2	9560
7	60	40	2.2	5.5	12,370
8	70	30	1.9	1.7	6530
9	90	10	1.4	5.6	10,440
	Petalite	**Kaolin**			
10	10	90	8.6	9.4	6600
11	20	80	7.9	11.0	11,400
12	30	70	5.7	12.6	15,230
13	40	60	4.7	10.8	11,680
14	50	50	4.2	9.1	9260
15	60	40			
16	70	30	−0.4	10.1	
17	90	10	0.6	10.3	8573

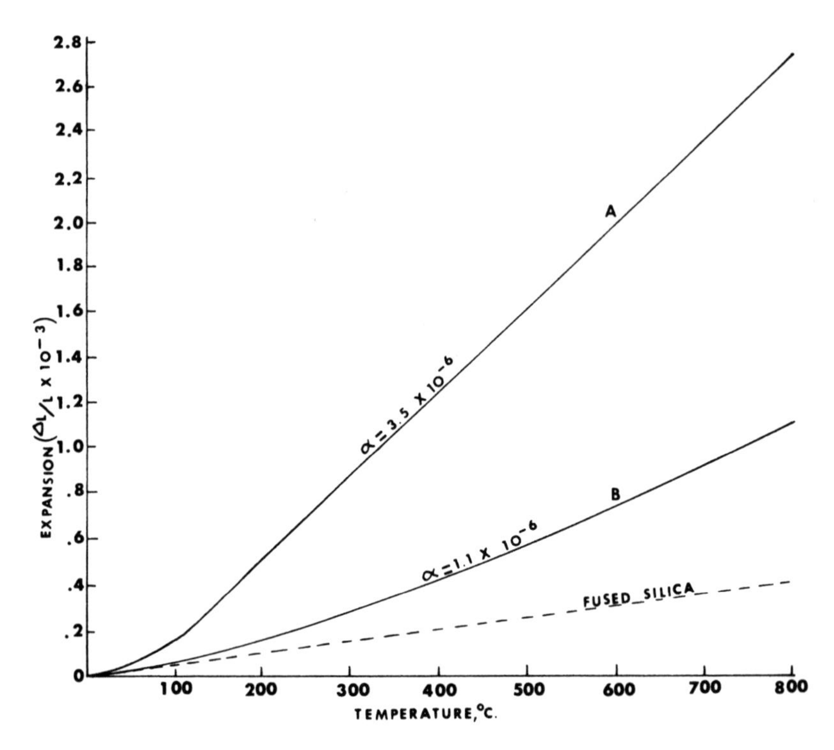

Fig. 15–4 Effect of firing temperature on resulting thermal expansion of a petalite-kaolin body. (A) 1310°C (2390°F) for 30 minutes, (B) 1290°C (2354°F) for 30 minutes.

Weltner[9] has produced true flameware bodies having compositions shown in Table 15–5.

All compositions when fired to Cone 11 have expansions below 2×10^{-6}/°C, and all survive repeated quenching in cold water after 10 minutes of dry heating on an electric hot plate.

Weltner, following the lead of Fishwick, reports that β-spodumene is a better replacement for the presently unavailable petalite than is α-spodumene. Alpha-spodumene undergoes expansion at 982°C (1800°F) resulting in a very porous Cone 11 body. Flux additions that overcome porosity lower thermal shock resistance.

Weltner[9] also reports that the lithium alumina-silicate bodies

TABLE 15–5
FLAMEWARE COMPOSITIONS[9]

	1	2	3	4
Petalite	30%	30%	–	
Spodumene (α)	20	20	30%	40% (β)*
Pyrophyllite	–	–	20	–
Talc	2	2	2	–
Fire clay	13	–	–	–
Jordan clay	15	27	27	–
Ball clay	18	20	20	60
Bentonite	2	1	1	–

* 10% 30 mesh, 30% 325 mesh

should be fired to Cone 11 or 12 in order to survive the previously described thermal shock test. The difference between Cones 10 and 11 is dramatic. Fired to Cone 10 all of the above bodies will fail in the thermal shock test, but at Cone 11 they will be completely satisfactory.

Low-Expansion Glazes

It is necessary to glaze a low-expansion body with a glaze having an even lower expansion in order to have a satisfactory glaze fit. Low-expansion bodies in the 0 to 3 $\times$ 10^{-6}/°C range present problems because conventional types of glaze compositions are not satisfactory.

The following range of glaze compositions have expansions of 4 $\times$ 10^{-6}/°C:

0.1 to 0.2 alkali		
0.4 PbO	0.4 Al_2O_3	5.0 SiO_2
0.4 to 0.5 CaO, MgO		1.3 to 1.7 B_2O_3

Duke[10] reports a composition having an expansion of 4.0×10^{-6}/°C, although it seems likely to have surface defects.

0.15 Na_2O		
0.065 K_2O	0.316 Al_2O_3	2.75 SiO_2
0.208 CaO		0.53 B_2O_3
0.712 PbO		

A glaze for cordierite bodies is reported in U.S. Patent 3,499,787 having the following composition:

Silica	43%
Alkali feldspar	10
Petalite	30
Alumina	7
Talc	5
Dolomite	5

Color: White   Surface: Matte   Expansion: 2.4×10^{-6}/°C

Thiess[4] reports that his cordierite bodies could be glazed by application of clay slips prepared from the same body plus the coloring oxides. The composition of the glaze slip used was as follows:

SiO_2	52.0%
Al_2O_3	13.5
MgO	15.5
B_2O_3	3.2
Fe_2O_3	9.0
Cr_2O_3	6.5

Talc, ball clay, boric acid and coloring oxides were used in formulating the above batch.

Smoke and Koenig[3] report a glaze that fits cordierite. This particular composition develops small amounts of barium magnesium alumina-silicate crystals that reduce the expansion of the glaze.

Oxide		**Batch**	
MgO	8.8%	Talc	25.1%
Al_2O_3	19.2	Clay	44.6
SiO_2	60.9	$BaCO_3$	12.6
BaO	11.1	Silica	17.7

Burnham and Tuttle[11] report that terra sigillata glazes fit talc-flint-calcined clay bodies, but little experimentation has been done with such glazes on the lower expansion bodies.

Probably the best approach to glaze development for the very low-expansion bodies involves the growth of low-expansion crystals from the glassy matrix, thereby reducing the overall expansion of the glaze.

Maki and Tashiro[8] have developed glazes applicable to lithia ceramic bodies. These glazes have expansion coefficients below $2.0 \times 10^{-6}/°C$ between room temperature and 500°C (932°F). The chemical composition of the glaze that showed the best results, having an expansion of $0.5 \times 10^{-6}/°C$, was as follows:

SiO_2	50.4%
Al_2O_3	29.2
Li_2O	5.9
ZrO_2	1.7
P_2O_5	2.6
TiO_2	2.6
Na_2O	1.0
K_2O	1.0
B_2O_3	2.8
PbO	2.8
	100.0

The above composition was fritted, then wet ball milled with 5 percent clay + 0.05 percent PVA (polyvinyl alcohol). It was screened through 200 mesh and applied to a previously bisqued body, dried and fired to 1290°C (2354°F) until the glaze melted and covered the surface (5 to 10 min.). It was then removed while at temperature from this furnace and placed in another furnace at a temperature of 750°C (1382°F). After 60 min. at 750°C it was removed from the furnace and cooled.

The second heat treatment grows crystals of β-eucryptite, $Li_2O{\cdot}Al_2O_3{\cdot}2SiO_2$, which has a negative expansion. The resulting glaze fits petalite bodies, has good chemical durability and excellent shock resistance from 500°C (932°F) to cold water.

Maki and Tashiro[8] point out the range of expansions available and the effect of the crystal growing temperature on the resulting expansion. Table 15–6 gives the composition and expansion of several low-expansion crystalline glazes.

The expansion curves for glazes 1, 2 and 3 of Table 15–6 are shown in Figure 15–5 together with a petalite-kaolin body and a fused silica for comparison.

Comparison of glazes 3, 4 and 5 of Table 15–6 indicates that control of the MgO content of these glazes may provide a good method of controlling glaze expansion in the 0.4 to 1.5×10^{-6}/°C range. The expansion of these glazes is plotted in Figure 15–6.

Weltner[9] reports a petalite glaze maturing at Cone 11 having the following composition:

Petalite	76.9%
Talc	14.2
Whiting	3.4
Kaolin	5.5

This glaze has a measured coefficient of expansion of 1×10^{-6}/°C. A glaze which fits the previously described Weltner[9] bodies can

TABLE 15–6
COMPOSITIONS OF SOME LOW-EXPANSION CRYSTALLINE GLAZES[8]

	1	2	3	4	5
SiO_2	44.8	44.8	50.4	50.4	50.4
Al_2O_3	30.0	34.2	29.2	29.2	29.2
Li_2O	10.7	6.4	5.9	5.9	5.9
ZrO_2	1.7	1.9	1.7	1.7	1.7
P_2O_5	2.6	2.8	2.6	2.6	2.6
TiO_2	2.6	2.6	2.6	2.6	2.6
Na_2O	1.2	1.0	1.0	1.0	1.0
K_2O	1.2	1.1	1.0	1.0	1.0
B_2O_3	2.6	2.6	2.8	2.8	2.8
PbO	2.6	2.6	2.8	2.8	2.8
MgO	—	—	—	2.0	4.0
	100.0	100.0	100.0	102.0	104.0
Expansion $\times 10^{-6}/°C$					
Reheat 750°C 1 hr.	1.5	0.5	0.2		
Reheat 800°C 1 hr.			0.4	1.0	1.5

be made by substituting α-spodumene for petalite and making small adjustments in the other ingredients. However, because of the 0.5 percent Fe_2O_3 in the spodumene, the resulting glaze will not be white.

In summary, one may say that there are ceramic bodies that have sufficiently low expansion coefficients to withstand any thermal shock normally encountered in the home. There are also glazes that will fit such bodies. If the potter wishes to make ware that is to undergo thermal shock, he should be prepared to pay close attention to com-

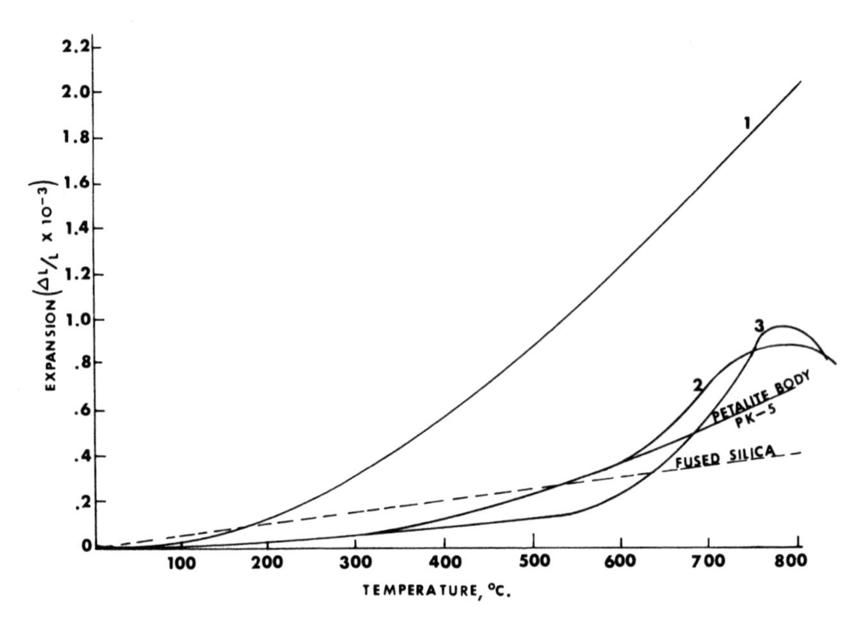

Fig. 15–5 Expansion curves for glazes 1, 2 and 3 of Table 15–6. Fused quartz and PK-5 body shown for comparison.

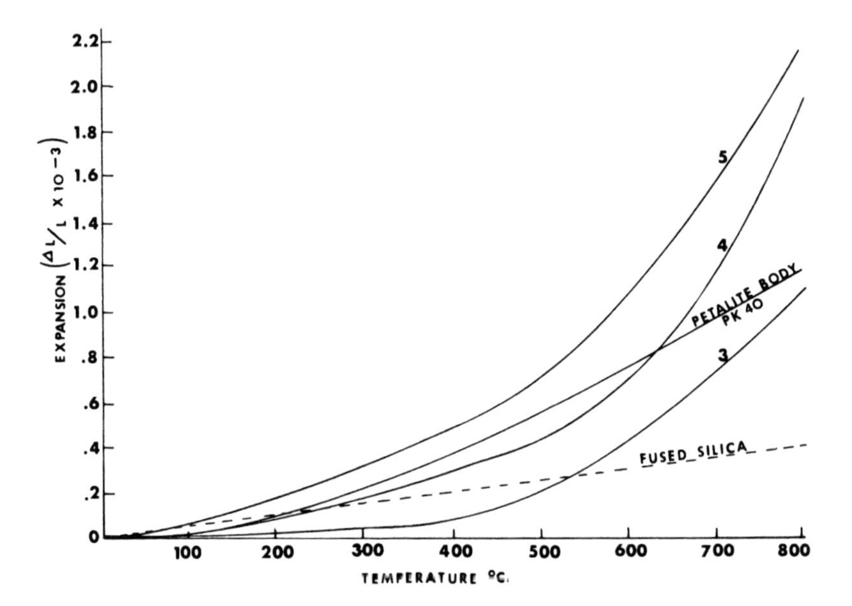

Fig. 15–6 Expansion curves for glazes 3, 4 and 5 of Table 15–6. This shows the effect of increasing amounts of MgO on resulting expansion properties of glaze.

positions, temperature control and duplication of time and temperature from one firing to another. Because of the great dependence of expansion properties on the previous heat treatment, slight differences can cause the difference between excellent and bad ware. Unless one is prepared to recognize these differences and cope with the control problems, one might better stay away from the sensitive bodies, such as cordierite and lithia ceramics. Properly manufactured, these compositions are most useful and beautiful and will withstand any thermal abuse in the home.

REFERENCES

1. Norton, F. H., *Fine Ceramics, Technology and Applications,* McGraw-Hill Book Co., Inc., N.Y., 1970.
2. Jelacic, C., "Practical Vitrified Cordierite Bodies for Porcelain Ovenware," *Ind. Ceram.,* 594: 213 (1967).
3. Smoke, E. J. and Koenig, J. H., "Thermal Properties of Ceramics," *Eng. Res. Bull.* No. 40, College of Engineering, Rutgers University, New Brunswick, N.J., 1958.
4. Thiess, L. E., "Vitrified Cordierite Bodies," *J. Am. Ceram. Soc.,* 26: 3 (1943).
5. Hummel, F. A., "Thermal Expansion Properties of Some Synthetic Lithia Minerals," *J. Am. Ceram. Soc.,* 34, 8: 235–39 (1951).
6. Anon., *Petalite,* Foote Mineral Co. Tech. Bull. No. 301, May, 1957.
7. Fishwick, J. H., Van Der Beck, R. R. and Talley, R. W., "Low Thermal Expansion Compositions in the System Spodumene-Kaolin and Petalite-Kaolin," *Bull. Am. Ceram. Soc.,* 43: 11 (1946).
8. Maki, T. and Tashiro, M., "Studies on the Glazes of Lithia Ceramics," *J. Ceram. Assoc. Japan,* 74, 3: 89 (1966).
9. Weltner, G., Personal Communication, May, 1971.
10. Duke, D. A., et al., "Strengthening Glass Ceramics by Application of Compressive Glazes," *J. Am. Ceram. Soc.,* 51: 98 (1968).
11. Burnham, F. and Tuttle, M. A., "Influence of Variable Amounts of New York Talc, Flint and Calcined Clay on Serviceability of Glazed Cooking Ware Bodies," *J. Am. Ceram. Soc.,* 28: 3 (1945).

CHAPTER 16

LEAD GLAZES, THEIR USE AND MISUSE

The use of lead compounds in glaze formulations imparts many desirable characteristics that account for their use in most commercial dinnerware glazes:[1]

1. **Low Melting Range.** The strong fluxing action of lead oxide, PbO, allows the formulation of glazes that mature at relatively low temperatures in comparison to their leadless counterparts.
2. **Wide Firing Range.** The PbO in the glaze reduces viscosity and allows for satisfactory maturing over a wider firing range. Such glazes are more foolproof.
3. **Low Surface Tension.** The PbO lowers the surface tension of a glaze, thereby contributing to the ability of the glaze to heal over blisters, drying cracks and other defects. It improves wetting, thus promoting adherence to the body and a smooth glossy surface.
4. **High Index of Refraction.** The PbO increases the index of refraction of a glass. Lead glazes are, therefore, more brilliant.

Occasionally there arises a case of lead poisoning apparently associated with the use of glazed ceramic ware. Whenever this happens, considerable publicity ensues with results that are damaging to the entire ceramic industry. Because of the national and international implications involved in lead release from glazed dinnerware, the Lead Industries Association, the U.S. Potters Association, the National Bureau of Standards and the International Lead Zinc Research Organization launched an extensive research effort to study the problem. This has resulted in a very comprehensive publication, *Lead Glazes for Dinnerware.*[1]

In reviewing instances in which lead poisoning has been determined as the cause of death, we find that, in most cases, it is the fault of the technically untrained potter.

About twenty years ago it was determined that lead poisoning had occurred after protracted use of an underfired dinnerware set. It was found that this glaze was an underfired matte glaze made with a frit designed to give a glossy texture at a higher firing temperature.

After this occurrence, the previously mentioned industrial and governmental groups made a thorough survey of commercial dinnerware glazes. It was found that all commercial dinnerware glazes tested were safe, except for the one offender. The offending glaze was found to be perfectly safe if fired to a higher temperature. Continuation of the survey into the fields of domestic artware, foreign commercial ware, foreign folkware and domestic hobby ware showed that some were indeed unsafe. Let us look at some further documented cases of lead poisoning to show where the offending ware originates.

In 1967, a glazed ceramic drinking vessel was confirmed as the source of lead poisoning.[2] The physician's report indicated that the victim habitually filled the container with a soft drink and sipped it during the evening. Often two bottles were consumed. After two years he was admitted to the hospital. The ceramic mug had been made by his son in a ceramic art course. Tests run on an unused mug from the same set revealed a lead release of over 300 ppm (7 ppm is considered the safe limit).[3]

Shortly thereafter, the well-publicized Hollyware case involving a Japanese import occurred. This ware had a green leaf and red berry design. The green was produced by the incorporation of copper oxide in the glaze. An alert ceramic salesman tested the lead release, which was found to be 100 ppm. U.S. Food and Drug Administration action resulted. Copper is known to result in high lead release when incorporated in lead glazes and should never be used.

Later a brown earthenware pitcher purchased in Mexico[4] and used for storing fruit juice claimed a victim, and another pitcher made by an amateur potter and purchased in Nova Scotia caused the death of a two-year-old boy who drank apple juice that had been stored in the pitcher.[5,6]

These cases illustrate the similarity of instances in which glazed pottery has been implicated in lead poisoning. The ceramic ware involved was, in most instances, manufactured by hobbyists or artware manufacturers ignorant of the factors involved in producing safe lead glazes.

It is hoped that the following information may help to assure the intelligent use of lead glazes on ware used as food or beverage containers. The word "intelligent" is stressed, and the following points are emphasized.[7]

1. "The presence of lead or other heavy metal compounds in a glaze does not, by itself, constitute a health hazard. The important factor is how resistant to attack by food acids the glaze may be."
2. "The most acid resistant of all ceramic glazes are lead-fluxed glazes, properly formulated, properly applied and properly fired."
3. "Commercially made dinnerware from the United States and a high percentage of that made abroad is well made and of very low lead release. (Most less than 1 ppm.)"
4. "Practically all ware made for food and beverage service, which has been involved in incidents of lead poisoning, comes from small studio potteries, hobby shops, and classrooms, or is homemade ware, foreign folkware or foreign specialty ware."
5. "Untrained people, adults or children, should be restrained from making glazed ceramic ware for food or beverage use. The potential harm is greatest when knowledge of the subject is minimal."

A ceramic glaze is a mixture of many oxides which fuse when heated to a high temperature, resulting in a glassy coating on the surface of the ware.

Even though ceramic materials are among the most permanent, nothing is completely insoluble and some glazes or glass compositions are more insoluble than others. The solubility of a fired glaze is measured by determining its acid resistance. Specifically, what is determined is the ability of a given concentration of acid to extract glaze ingredients from a sample under rigid test conditions. Using these tests, it is possible to determine the amount of lead, cadmium, copper, barium or other potentially toxic heavy metals that leach from a glaze.

If the glaze has been properly formulated, applied and fired, the quantity of any of these is very small. For lead, it is less than 7 ppm—an amount scarcely measurable with accuracy less than twenty years ago. Normally, these extraction tests use an organic acid similar to vinegar, acetic acid.

The best current test is that developed by the U.S. Food and Drug Administration, Division of Compliance Programs, Bureau of Foods. This method was originally published as FDA Laboratory Information Bulletin No. 834, and is reproduced in the Appendix along with an older test for lead release, ASTM C-555-71.

The FDA test is used for product safety enforcement, and has gained universal application in the United States. It has replaced the older ASTM test. *All* food service utensils manufactured in the United States must meet the FDA lead release guideline of less than 7 micrograms of lead per milliliter (μg./ml.) of leaching solution when analyzed by the FDA method.

As noted earlier, all glazes are soluble in acid to some extent. Research by the International Lead Zinc Research Organization and the U.S. Potters Association has shown that certain processing variables within the formulation, application and firing operations have major influence on the resultant acid resistance and, in the absence of proper control, can greatly increase the solubility of lead and other heavy metals from the glaze. It is this lack of process control which produces high-solubility glazes that are cause for concern.

Poorly designed glazes, underfired glazes and particularly poorly formulated glazes maturing in the Cone 07 to 04 range augment the lead solubility problem. Other factors, such as failure to fully frit the glaze, the presence of copper oxide, application thickness and firing conditions also affect lead release. It is important to note that it is *not* the percentage of lead present in the glaze that determines the amount of lead solubility, but rather the ratio of the glaze ingredients, the time and the temperature of firing.

Control of Lead Release

Control of lead release is essential if the hobbyist's or potter's end product is to present no danger to the public with respect to lead solubility.

The research program dealing with the variables effecting lead release from glazes points out the important variables to be considered by people working with lead glazes. The criteria for designing acid resistant lead glazes are based on composition, structure and fabrication techniques. Their understanding will result in intelligent application of lead glazes and will eliminate any apprehension on the part of the potter and the public.

Early work was done in England by Thorpe,[8] who arrived at an empirical relationship that predicted the solubility of lead frits. This is known as Thorpe's ratio:

$$\frac{\text{Moles of Basic Oxides} + \text{Moles Alumina}}{\text{Moles of Acidic Oxides}} \times \frac{223}{60} = \text{Thorpe's ratio}$$

This ratio should not exceed 2.0 for low solubility. Mellor[9] later restated it as follows:

$$\frac{RO + Al_2O_3}{RO_2} = 0.5 \text{ maximum.}$$

Thus, for a glaze having the following composition:

0.5 PbO		
0.3 Na_2O	0.15 Al_2O_3	1.6 SiO_2,
0.2 K_2O		

Thorpe's ratio is

$$\frac{1 + 0.15}{1.6} \times \frac{223}{60} = 2.7,$$

which is above the acceptable limit of 2.0. Mellor's ratio is

$$\frac{1 + 0.15}{1.6} = 0.71,$$

which is above the acceptable limit of 0.5.

More recently, Eppler[10] has refined the technique of predicting lead release. His technique is based on the fact that lead release is not directly related to the concentration of lead oxide in a glaze, but rather is influenced by the other constituents.

It has been shown that silica, alumina, zirconia, titania and tin oxide are effective in lowering the lead release of a glaze. These oxides are considered good additives.

$$\text{Good} = 2\ Al_2O_3 + SiO_2 + ZrO_2 + TiO_2 + SnO_2 \quad (1)$$

The concentrations are expressed in the empirical formula concentrations conventionally used in the whiteware industry. The number 2 arises from the fact there are two equivalents of aluminum ions in each equivalent of Al_2O_3.

It has also been shown that alkalies, alkaline earths, boron oxide, fluorine, phosphate, zinc oxide, cadmium oxide and lead oxide are all more or less effective in increasing lead release from a glaze.

$$\text{Bad} = 2(Li_2O + Na_2O + K_2O + B_2O_3 + P_2O_5) + MgO + CaO + SrO + BaO + F + ZnO + CdO + PbO \quad (2)$$

Again the number 2 appears in order to relate the oxide concentration to the ionic equivalents produced. These ions consist of the soluble network former B_2O_3, which disrupt the silicate network and the network modifiers.

Applying Equations 1 and 2 to 77 glazes using computer-assisted regression analysis, a figure of merit calculation was developed to predict lead release. The best method of calculating the figure of merit was to divide the "good" factor (Equation 1) by the square root of the "bad" factor (Equation 2).

$$\text{Figure of Merit} = \frac{\text{Good}}{\sqrt{\text{Bad}}} \quad (3)$$

If the figure of merit exceeded 2.05, the lead release was less than 7 ppm. If less than 1.80 some measurements of lead release exceeded 7 ppm, although individual readings were less. The figure of merit could not discriminate acceptability of lead release in the 1.80 to 2.05 range.

This figure of merit has been found applicable to all gloss glazes but not to matte or crystalline glazes in which crystalline phases are precipitated from the glass.

These empirical relationships may be used only as rough guides to the eventual solubility of the glaze on the finished ware. This is the result of the many variables entering the formulation, application and firing. These variables are summarized in the ILZRO manual[1] and briefly discussed here.

1. Cone 3 to 5 clear lead glazes show lead release values of less than 0.5 ppm and, in many cases, less than 0.1 ppm. This constitutes the major portion of all commercial dinnerware glazes used in the United States. A typical composition of this type of glaze is as follows:

0.066 K_2O		
0.179 Na_2O	0.340 Al_2O_3	3.369 SiO_2
0.261 PbO	0.314 B_2O_3	
0.494 CaO		

   All of the glaze is fritted except the necessary mole equivalents of Al_2O_3 and SiO_2 to provide a 10 percent mill addition. The average lead release from this glaze is 0.16 ppm.
2. The effect of various coloring oxides and commercial stains on lead release was studied. The coloring oxides included CuO, Cr_2O_3, Co_3O_4, Fe_2O_3, MnO_2, Fe_2O_3-MnO_2, $ZrO_2 \cdot SiO_2$, PbSb yellow stain, Cr-Al pink stain, Sn-Sb gray stain and Co-Cr-Fe black stain. These additions were made to various lead glazes having maturing ranges from Cone 4 to 05. In all cases except one, CuO, no effect on lead release was found and all glazes were well below the acceptable level.

3. A 2 percent addition of CuO to a low-temperature lead glaze maturing at 1024°C (1875°F) resulted in a lead release increase *ten times that of the base glaze.* It is necessary to state emphatically that *copper oxide should not be used* either as a direct addition or component of a stain in a lead glaze. Although this effect led to the abandonment years ago of the copper green glaze by industry, it should be forcibly brought to the attention of the hobby and art potter.
4. The addition of opacifiers, such as zircon, tin oxide or titanium oxide, does not increase lead release. In fact, in some cases it decreases lead release.
5. For a Cone 4 lead glaze, variation in the R_2O and RO content showed little effect. The same observations were made when these glazes were fired to Cone 01. Additions of B_2O_3 does increase the lead release from these glazes.
6. In low-temperature (Cone 07) lead silicate glazes, $PbO{\cdot}1.3SiO_2$, small, single alkali additions increase the lead release with Li_2O increasing it the least, Na_2O intermediate and K_2O increasing it the most. As atomic size increases $Li^+ \rightarrow Na^+ \rightarrow K^+$, the lead release also increases. A mixture of any two of the three alkali oxides totaling 0.1 mole equivalents results in lower lead release than the same mole equivalent addition of a single oxide with a mixture of all three giving the best results.

   The alkaline earth oxides gave similar results with lead release increasing from $Mg^{+2} \rightarrow Ca^{+2} \rightarrow Sr^{+2}$. Again best results were obtained by using mixtures of the three oxides.
7. In a low-temperature (Cone 03) lead silicate, $PbO{\cdot}1.5SiO_2$, base glaze, Al_2O_3 additions reduced lead release and B_2O_3 additions increased lead release. However, in a Cone 1 glaze, these results could not be duplicated.
8. In practically all cases, the amount of lead leached from the glaze surface decreases with repeated testing. On a Cone 4 to 5 glaze, this dropped from 0.10 ppm on the first test to 0.06 ppm on the eighth test. On a Cone 05 glaze with 2 percent CuO added, it dropped from 0.78 ppm to 0.12 ppm. This is probably due to the fact that lead, being a surface tension reducer,

concentrates on the surface, and, therefore, the initial leaching lowers the concentration available for subsequent removal.

9. The greater the thickness of the glaze application, the larger the lead release. For Cone 4 to 5 glazes, the effect of thickness is insignificant, but, for a Cone 07 low-temperature glaze, this becomes an important factor.
10. Longer firing times and higher temperatures result in decreased lead release. This is probably due to increased glaze-body reaction and more solution of the body into the glaze, thereby, changing its composition.
11. Marquis[11] has called attention to the effect of firing conditions on lead release, particularly stagnant versus moving air. No such effect was noted on firing the Cone 3 to 5 lead glazes, and they always produce a very low lead release value even though they contain 15 to 20 percent PbO. Glazes maturing at Cone 06, which may contain more than 20 percent PbO, as well as other volatile constituents such as CaF_2, often give erratic results. They are sensitive to application thickness as well as movement of the atmosphere during firing. When fired in a gas-fired kiln, the test results are consistently low with little variation. When fired in an electric kiln with a static atmosphere, test results on a particular glaze varied from 0.07 ppm to 24.8 ppm with an average of 6.8 ppm. If the electric kiln is vented to provide air circulation, consistently low results are obtained.

The following pointers should help the individual potter to eliminate lead release problems:

1. There is no problem with Cone 3 to 5 lead glazes when properly compounded and fritted.
2. In preparing low temperature lead glazes, use mixtures of R_2O and RO oxides rather than single components.
3. Al_2O_3 additions decrease lead release, while B_2O_3 additions increase lead release.
4. Never use copper in any form in a lead glaze.
5. Use thin glaze applications.

6. Fire in a circulating atmosphere. If electric kilns are used, provide vents for air movement.
7. Check lead release from the finished ware until assured of its safe level. For details of the Glaze Surveillance program and lead release tests see Appendix.
8. See Appendix for Cone 06 glazes proven satisfactory for electric kiln firing.

REFERENCES

1. *Lead Glazes for Dinnerware,* International Lead Zinc Research Organization, Inc., N.Y., 1971.
2. Harris, R. W. and Elsea, W. R., "Ceramic Glaze as a Source of Lead Poisoning," *J. Am. Medical Assoc.,* 202: 6 (1967).
3. Smith, Jerome F., "Role of Lead Industries Association and International Lead Zinc Research Organization in the Program to Control Lead Release from Glaze Surfaces Intended for Contact with Food," Proceedings, Fall 1970 Meeting, Materials and Equipment and Whitewares Divisions, American Ceramic Society.
4. Block, J. L., "The Accident That Saved Five Lives," p. 63, *Good Housekeeping,* November, 1969.
5. Bieler, Z., "Death Can Lurk in an Earthenware Pot," p. 4, *The Montreal Star,* December 27, 1969.
6. Cole, Jerome F., "Health Aspects of ILZRO's Lead in Glaze Research," Proceedings, Fall 1970 Meeting, Materials and Equipment and Whitewares Divisions, American Ceramic Society.
7. Nordyke, J. S., Private Communication, March 9, 1971.
8. Thorpe, T. E., "The Use of Lead in the Manufacture of Pottery," 1899, Government Paper 8383-1500931/1901; also "Report on Work of the Government Laboratory on the Question of Employment of Lead Compounds in Pottery," 1901, Government Paper 9264-1500-61901; "Lead in the Potteries," London, 1910.
9. Mellor, J. W., "The Durability of Pottery Frits, Glazes and Enamels in Service," *Trans. Brit. Ceram. Soc.,* 34: 113–90 (1934).

10. Eppler, R. A., "Formulation and Processing of Ceramic Glazes for Low Lead Release," Proc. International Conference on Ceramic Foodware Safety, Geneva, Switzerland, November 12–14, 1974.
11. Marquis, J. E., "Lead in Glazes—Benefits and Safety Precautions," Proceedings, Fall 1970 Meeting, Materials and Equipment and Whitewares Divisions, American Ceramic Society.

A P P E N D I C E S

APPENDIX A–1

TEMPERATURE EQUIVALENTS OF ORTON PYROMETRIC CONES*

Cone No.	20°C/Hr. °C	20°C/Hr. °F	150°C/Hr. °C	150°C/Hr. °F	Cone No.	20°C/Hr. °C	20°C/Hr. °F	150°C/Hr. °C	150°C/Hr. °F
022	585	1090	605	1120	01	1110	2030	1145	2090
021	595	1100	615	1140	1	1125	2060	1160	2120
020	625	1160	650	1200	2	1135	2080	1165	2130
019	630	1170	660	1220	3	1145	2090	1170	2140
018	670	1240	720	1330	4	1165	2130	1190	2170
017	720	1330	770	1420	5	1180	2160	1205	2200
016	735	1360	795	1460	6	1190	2170	1230	2250
015	770	1420	805	1480	7	1210	2210	1250	2280
014	795	1460	830	1530	8	1225	2240	1260	2300
013	825	1520	860	1580	9	1250	2280	1285	2350
012	840	1540	875	1610	10	1260	2300	1305	2380
010	890	1630	905	1640	11	1285	2350	1325	2420
09	930	1710	930	1710	12	1310	2390	1335	2440
08	945	1730	950	1740	13	1350	2460	1350	2460
07	975	1790	990	1810	14	1390	2530	1400	2550
06	1005	1840	1015	1860	15	1410	2570	1435	2560
05	1030	1890	1040	1900	16	1450	2640	1465	2670
04	1050	1920	1060	1940	17	1465	2670	1475	2690
03	1080	1980	1115	2040	18	1485	2710	1490	2720
02	1095	2000	1125	2060					

The higher temperature cones manufactured by the Orton Foundation are omitted from this listing.

*Fairchild, C. O. and Peters, M. F., *J. Am. Ceram. Soc.* 1, 701 (1926)

APPENDIX A–2

TABLE OF ATOMIC WEIGHTS

Element	Symbol	Atomic No.	Atomic Weight	Element	Symbol	Atomic No.	Atomic Weight
Actinium	Ac	89	227.0	Hafnium	Hf	72	178.5
Aluminum	Al	13	26.9	Helium	He	2	4.0
Americium	Am	95	243.0	Holmium	Ho	67	164.9
Antimony	Sb	51	121.8	Hydrogen	H	1	1.0
Argon	A	18	39.9	Indium	In	49	114.8
Arsenic	As	33	74.9	Iodine	I	53	126.9
Astatine	At	85	210.0	Iridium	Ir	77	192.2
Barium	Ba	56	137.4	Iron	Fe	26	55.9
Beryllium	Be	4	9.0	Krypton	Kr	36	83.8
Bismuth	Bi	83	209.0	Lanthanum	La	57	138.9
Boron	B	5	10.8	Lead	Pb	82	207.2
Bromine	Br	35	79.9	Lithium	Li	3	6.9
Cadmium	Cd	48	112.4	Lutetium	Lu	71	175.0
Calcium	Ca	20	40.1	Magnesium	Mg	12	24.3
Carbon	C	6	12.0	Manganese	Mn	25	54.9
Cerium	Ce	58	140.1	Mercury	Hg	80	200.6
Cesium	Cs	55	132.9	Molybdenum	Mo	42	96.0
Chlorine	Cl	17	35.5	Neodymium	Nd	60	144.3
Chromium	Cr	24	52.0	Neon	Ne	10	20.2
Cobalt	Co	27	58.9	Neptunium	Np	93	237.0
Copper	Cu	29	63.5	Nickel	Ni	28	58.7
Curium	Cm	96	248.0	Niobium	Nb	41	92.9
Dysprosium	Dy	66	162.5	Nitrogen	N	7	14.0
Erbium	Er	68	167.3	Osmium	Os	76	190.2
Europium	Eu	63	152.0	Oxygen	O	8	16.0
Fluorine	F	9	19.0	Palladium	Pd	46	106.4
Francium	Fr	87	223.0	Phosphorous	P	15	31.0
Gadolinium	Gd	64	157.3	Platinum	Pt	78	195.1
Gallium	Ga	31	69.7	Plutonium	Pu	94	244.0
Germanium	Ge	32	72.6	Polonium	Po	84	210.0
Gold	Au	79	197.0	Potassium	K	19	39.1

Element	Symbol	Atomic No.	Atomic Weight	Element	Symbol	Atomic No.	Atomic Weight
Praseodymium	Pr	59	140.9	Tantalum	Ta	73	180.9
Promethium	Pm	61	145.0	Technetium	Tc	43	99.0
Protactinium	Pa	91	231.0	Tellurium	Te	52	127.6
Radium	Ra	88	226.1	Terbium	Tb	65	159.0
Radon	Rn	86	222.0	Thallium	Tl	81	204.4
Rhenium	Re	75	186.2	Thorium	Th	90	232.1
Rhodium	Rh	45	102.9	Thulium	Tm	69	169.0
Rubidium	Rb	37	85.5	Tin	Sn	50	118.7
Ruthenium	Ru	44	101.1	Titanium	Ti	22	47.9
Samarium	Sm	62	150.4	Tungsten	W	74	183.9
Scandium	Sc	21	45.0	Uranium	U	92	238.1
Selenium	Se	34	79.0	Vanadium	V	23	51.0
Silicon	Si	14	28.1	Xenon	Xe	54	131.3
Silver	Ag	47	107.9	Ytterbium	Yb	70	173.0
Sodium	Na	11	23.0	Yttrium	Y	39	88.9
Strontium	Sr	38	87.6	Zinc	Zn	30	65.4
Sulfur	S	16	32.1	Zirconium	Zr	40	91.2

APPENDIX A–3

EQUIVALENT WEIGHTS OF SOME COMMON CERAMIC MATERIALS

Material	Formula	Molecular Weight	Equivalent Weight RO R_2O	R_2O_3	RO_2
Aluminum Oxide	Al_2O_3				
Alumina hydrate	$Al_2O_3 \cdot 3H_2O$	156		156	
Anorthite	$CaO \cdot Al_2O_3 \cdot 2SiO_2$	279	279	279	139
Clays, kaolins	$Al_2O_3 \cdot 2SiO_2 \cdot 2H_2O$	258		258	129
Clays, calcined	$Al_2O_3 \cdot 2SiO_2$	222		222	111
Cornwall stone	See Potassium Oxide				
Cryolite	Na_3AlF_6	210	140	420	
Feldspar, potash orthoclase or microcline	$K_2O \cdot Al_2O_3 \cdot 6SiO_2$	557	557	557	93
Feldspar, soda albite	$Na_2O \cdot Al_2O_3 \cdot 6SiO_2$	525	525	525	88
Kyanite	$Al_2O_3 \cdot SiO_2$	162		162	162
Plastic vitrox	$.053CaO$ $.334Na_2O$ $1.33Al_2O_3$ $.613K_2O$ $13.95SiO_2$	1051	1051	790	76
Pyrophyllite	$Al_2O_3 \cdot 4SiO_2 \cdot H_2O$	360		360	90
Antimony Oxide	Sb_2O_3	292		292	
Barium Oxide	BaO				
Barium carbonate or witherite	$BaCO_3$	197	197		
Barytes	$BaSO_4$	233	233		
Boric Oxide	B_2O_3				
Borax	$Na_2O \cdot 2B_2O_3 \cdot 10H_2O$	382	382	191	

Material	Formula	Molecular Weight	Equivalent Weight RO R_2O	R_2O_3	RO_2
Borax, calcined to					
62°C	$Na_2O \cdot 2B_2O_3 \cdot 5H_2O$	292	292	146	
130°C	$Na_2O \cdot 2B_2O_3 \cdot 3H_2O$	256	256	128	
180°C	$Na_2O \cdot 2B_2O_3 \cdot H_2O$	220	220	110	
318°C	$Na_2O \cdot 2B_2O_3$	202	202	101	
Boric acid	$B_2O_3 \cdot 3H_2O$	124		124	
Razorite	$Na_2O \cdot 2B_2O_3 \cdot 4H_2O$	274	274	137	
Gerstley borate	$Na_2O \cdot 2CaO \cdot 5B_2O_3 \cdot 8H_2O$	652	217	130	
Colemanite	$2CaO \cdot 3B_2O_3 \cdot 5H_2O$	412	206	137	
Cadmium Oxide	CdO				
Cadmium carbonate	$CdCO_3$	172	172		
Calcium Oxide	CaO				
Anorthite	$CaO \cdot Al_2O_3 \cdot 2SiO_2$	279	279	279	139
Bone ash	$3CaO \cdot P_2O_5$	310	103	310	
Colemanite	$2CaO \cdot 3B_2O_3 \cdot 5H_2O$	412	206	137	
Calcite (whiting)	$CaCO_3$	100	100		
Dolomite	$CaCO_3 \cdot MgCO_3$	184	92		
Fluorspar	CaF_2	78	78		
Wollastonite	$CaO \cdot SiO_2$	116	116		116
Chromium Oxide	Cr_2O_3				
Barium chromate	$BaCrO_4$	253	253	506	
Lead chromate	$PbCrO_4$	323	323	646	
Potassium chromate	K_2CrO_4	194	194	388	
Cobalt Oxide	$CoO \cdot Co_3O_4$				
Cobalt oxide (black)	Co_3O_4	241	80		
Cobalt oxide (gray)	CoO	75	75		
Cobalt carbonate	$CoCO_3$	119	119		
Copper Oxide	CuO				
Copper carbonate	$2CuCO_3 \cdot Cu(OH)_2$	345	115		

Material	Formula	Molecular Weight	Equivalent Weight		
			RO R_2O	R_2O_3	RO_2
Copper oxalate	$CuC_2O_4 \cdot \frac{1}{2}H_2O$	161	161		
Cupric oxide (black)	CuO	80	80		
Cuprous oxide (red)	Cu_2O	144	72		
Iron Oxide	Fe_2O_3				
Ferrous oxide (black)	FeO	72	72		
Ferric oxide (red)	Fe_2O_3	160	80	160	
Magnetite	$FeO \cdot Fe_2O_3$	231	116	231	
Siderite	$FeCO_3$	116	116		
Lead Oxide	PbO				
Litharge	PbO	223	223		
Lead monosilicate	$PbO \cdot 0.67SiO_2$	263	263	393	
Lead bisilicate	$PbO \cdot 0.03Al_2O_3 \cdot 1.95SiO_2$	344	344		176
Lead bisilicate	$PbO \cdot 0.25Al_2O_3 \cdot 1.91SiO_2$	364	364		191
Red lead	Pb_3O_4	686	229		
White lead	$2PbCO_3 \cdot Pb(OH)_2$	776	259		
Tribasic lead silicate	$PbO \cdot 0.33SiO_2$	243	243		736
Lithium Oxide	Li_2O				
Lithium carbonate	Li_2CO_3	74	74		
Spodumene	$Li_2O \cdot Al_2O_3 \cdot 4SiO_2$	372	372	372	93
Petalite	$Li_2O \cdot Al_2O_3 \cdot 8SiO_2$	612	612	612	76
Magnesium Oxide	MgO				
Brucite	$Mg(OH)_2$	58	58		
Dolomite	$CaCO_3 \cdot MgCO_3$	184	92		
Magnesium carbonate	$MgCO_3$	84	84		
Magnesia	MgO	40	40		
Talc	$3MgO \cdot 4SiO_2 \cdot H_2O$	378	126		95
Serpentine	$3MgO \cdot 2SiO_2 \cdot 2H_2O$	276	92		188
Manganese Oxide	MnO				
Manganese carbonate	$MnCO_3$	115	115		
Manganese dioxide	MnO_2	87	87		
Manganese oxide	MnO	71	71		

Material	Formula	Molecular Weight	Equivalent Weight RO R_2O	R_2O_3	RO_2
Nickel Oxide	NiO				
Nickel carbonate	$NiCO_3$	119	119		
Nickel oxide (green)	NiO	75	75		
Nickel oxide (black)	Ni_2O_3	166	83		
Phosphorous Pentoxide	P_2O_5				
Bone ash	$3CaO{\cdot}P_2O_5$	320	107		320
Potassium Oxide	K_2O				
Cornwall stone	This is a mixture of quartz, mica and kaolin with the following typical composition: SiO_2 72.5%, Al_2O_3 16.5%, Fe_2O_3 0.2%, TiO_2 0.1%, CaO 2.0%, MgO 0.2%, Na_2O 3.5% and K_2O 4.0%.				
Nephelene syenite	$Na_2O{\cdot}Al_2O_3{\cdot}2SiO_2$	284	284	284	142
Feldspar, potash microcline or orthoclase	$K_2O{\cdot}Al_2O_3{\cdot}6SiO_2$	557	557	557	93
Pearl ash	K_2CO_3	138	138		
Plastic vitrox	.053 CaO 1.33 Al_2O_3 .334 Na_2O 13.9 SiO_2 .613 K_2O	1051	1051	790	75.6
Silicon Dioxide	SiO_2				
Clays, kaolins	$Al_2O_3{\cdot}2SiO_2{\cdot}2H_2O$	258		258	129
Clays, calcined	$Al_2O_3{\cdot}2SiO_2$	222		222	111
Cornwall stone	See Potassium Oxide.				
Feldspar, potash	$K_2O{\cdot}Al_2O_3{\cdot}6SiO_2$	557	557	557	93
Feldspar, soda	$Na_2O{\cdot}Al_2O_3{\cdot}6SiO_2$	524	524	524	87
Nephelene syenite	$Na_2O{\cdot}Al_2O_3{\cdot}2SiO_2$	284	284	284	142
Flint or quartz	SiO_2	60			60
Plastic vitrox	See Potassium Oxide.				
Pyrophyllite	$Al_2O_3{\cdot}4SiO_2{\cdot}H_2O$	360		360	90

Material	Formula	Molecular Weight	Equivalent Weight RO R_2O	R_2O_3	RO_2
Talc	$3MgO \cdot 4SiO_2 \cdot H_2O$	379	126		95
Wollastonite	$CaO \cdot SiO_2$	116	116		116
Sodium Oxide	Na_2O				
Borax	See Boric Oxide.				
Borax, calcined					
Razorite					
Feldspar, soda	See Silicon Dioxide.				
Nephelene syenite	See Silicon Dioxide.				
Cryolite	See Aluminum Oxide.				
Plastic vitrox	See Potassium Oxide.				
Strontium Oxide	SrO				
Strontium carbonate	$SrCO_3$	148	148		
Tin Oxide	SnO_2				
Tin oxide	SnO_2	151			151
Titanium Oxide	TiO_2				
Illmenite	$FeO \cdot TiO_2$	152	152		152
Lead titanate	$PbO \cdot TiO_2$	303	303		303
Rutile	TiO_2	80			80
Zinc Oxide	ZnO				
Zinc oxide	ZnO	81	81		
Zirconium Oxide	ZrO_2				
Zirconium oxide	ZrO_2	123			123
Zircon or zirconium Silicate	$ZrO_2 \cdot SiO_2$	183			183

APPENDIX A–4

RAPID SCREENING METHOD FOR LEAD RELEASED BY CERAMIC GLAZES*

From: The Kettering Laboratory
Department of Preventive Medicine and Industrial Health
College of Medicine
University of Cincinnati
Cincinnati, Ohio

1. Principle:

A kit has been developed for the rapid determination of lead in 5 percent acetic acid used in testing ceramic glazes. A single extraction with dithizone is made in a modification of equipment developed by members of the technical staff of Ethyl Corporation for the determination of organolead in air. This procedure will not eliminate the interference due to bismuth.

2. Apparatus:

a. Special color comparator tube providing a light path of 1 cm. and with a capacity of approximately 100 ml.
b. Hellige comparator standardized for the estimation of lead in ppm.
c. 16 ml. pipette.

3. Reagents:

Distilled Water
Vial A—30 ml. of a buffer solution prepared from:
Two g. hydroxylamine hydrochloride
Twenty g. ammonium citrate

*Reproduced by permission of International Lead Zinc Research Organization, 292 Madison Avenue, New York, N.Y. 10017.

Dissolve the above reagents in 300 ml. of distilled water and make the solution alkaline to phenol red indicator by the addition of reagent grade ammonium hydroxide. Add 10 g. of potassium cyanide and 1950 ml. of ammonium hydroxide (Sp. Gr. 0.9) to complete the buffer solution.

Vial B—15 ml. chloroform conditioned as follows:

Introduce 1 L. of chloroform, redistilled from a borosilicate glass still, into a 2 L. glass-stoppered borosilicate glass separatory funnel. Dissolve approximately 10 g. of hydroxylamine hydrochloride in 50 ml. of distilled water and make the solution alkaline to phenol red indicator by the addition of reagent grade ammonium hydroxide. Add this solution to the chloroform in the separatory funnel and shake well. Allow the aqueous layer to separate and filter the chloroform through a fluted filter paper into a brown, glass-stoppered bottle containing 20 ml. of absolute ethyl alcohol. Shake well and store in refrigerator.

Vial C—Dithizone

0.3 mg. of dithizone dissolved in 1 ml. of conditioned chloroform and evaporated to dryness.

These three solutions are described in detail for those who wish to prepare their own reagents. The reagents and vials can be provided by a reputable chemical supply house.

Procedure

1. Take, at random, 3 cups, and cleanse each with a Calgonite rinse (1 tablespoon per gallon of lukewarm water).
2. Heat 450 ml. of 5 percent acetic acid to 60°C (140°F) and introduce 150 ml. into each of the 3 cups. Let stand for 30 minutes.
3. Combine the acetic acid from the 3 cups by transferring the solutions to a 500 ml. volumetric flask. Mix well.
4. Introduce 16 ml. of acetic acid solution by pipette into the special color comparator tube.
5. Add 35 ml. of distilled water and mix by shaking.
6. Break an ampoule containing solution A, and add its contents to the color comparator tube. Mix well.

7. Prepare the dithizone solution by adding the chloroform in ampoule B to the vial containing the dry dithizone (ampoule C). Add the dithizone solution to the comparator tube, and shake vigorously for 1 min.
8. Insert the color comparator tube into the Hellige comparator, and match the color with that of the correct standard glass disc of the comparator. The numbers in the upper right corners are read directly as lead in ppm in the aliquot of 16 ml.

Note

This single extraction procedure can be used with a photoelectric filter photometer or spectrophotometer to yield a more precise estimate of the lead content of the solution than that provided by the colored disc comparator. If the colored disc comparator is not used, a working graph may be prepared by adding known quantities of lead to 16 ml. of 5 percent acetic acid, and following the analytical procedure after adjusting the volume to 50 ml. with distilled water.

APPENDIX A–5

DETERMINATION OF LEAD IN POTTERY

FDA-LIB-834

Apparatus and Reagents

(a) Atomic absorption spectrometer—Perkin-Elmer Model 303 or equivalent, with the following operating conditions: wavelength 2180Å.; slit 4; lead hollow cathode lamp; air acetylene burner (0.5 × 110 mm. slit) with supply of air at 60 psi (flow meter 9.0) and acetylene at 10 psi (flow meter 9.0) for an aspiration rate of 0.8 ml./min.

(b) Standard solution—Dissolve any soluble lead salt in 4 percent acetic acid to a lead concentration of 1 mg./ml. Dilute this standard stock solution with 4 percent acetic acid to obtain working standards with final concentrations of 10, 20, 30 and 40 μg. lead/ml.

Preparation of Sample (Leaching) Solution

(Individually analyze 3 units of each sample)

Prior to analysis, wash all vessels with household detergent, followed by a thorough rinse with distilled water. Determine and record the normal capacity of one unit by pouring a measured volume of distilled water into the vessel to a level to which it would normally be filled by the user. Discard the water and dry the unit; then fill each unit with 4 percent acetic acid to a volume determined by the preceding step. Cover with a watch glass or other suitable cover and let stand at room temperature (25°C ± 2°C) for 24 hours.

Determination

Stir sample (leaching) solution and determine absorbance by atomic absorption spectrometry, diluting if required with 4 percent acetic acid. Determine the absorbance of the standard solutions in a similar fashion.

Prepare a standard curve of absorbance versus concentration. Determine the amount of lead from the standard curve. Calculate results as μg. lead/ml. of leaching solution.

A sample is considered violative if the average of the 3 units examined contains 7.0 μg. lead/ml. of leaching solution or more.

APPENDIX A–6

DETERMINATION OF CADMIUM IN POTTERY

Apparatus and Reagents

(a) Atomic absorption spectrometer—Perkin-Elmer Model 303 or equivalent, with the following operating conditions: wavelength 2288Å.; slit 4; cadmium hollow cathode lamp; air acetylene burner (0.5 × 110 mm. slit) with supply of air at 60 psi (flow meter 9.0) and acetylene at 10 psi (flow meter 9.0) for an aspiration rate of 0.8 ml./min.

(b) Standard solution—Dissolve any soluble cadmium salt in 4 percent acetic acid to a cadmium concentration of 0.1 mg./ml. Dilute this standard stock solution with 4 percent acetic acid to obtain working standards with final concentrations of 1, 2, 3 and 4 μg. cadmium/ml.

Preparation of Sample (Leaching) Solution

(Individually analyze 3 units of each sample)

Prior to analysis, wash all vessels with household detergent, followed by a thorough rinse with distilled water. Determine and record the normal capacity of one unit by pouring a measured volume of distilled water into the vessel to the level to which it would normally be filled by the user. Discard the water and dry the unit; then fill each unit with 4 percent acetic acid to the volume determined in the preceding step. Cover with a watch glass or other suitable cover, and let stand at room temperature (25°C ± 2°C) for 24 hours.

Determination

Stir sample (leaching) solution and determine absorbance by atomic absorption spectrometry, diluting if required with 4 percent acetic acid. Determine the absorbance of the standard solutions in a similar fashion.

Prepare a standard curve of absorbance versus concentration. Determine the amount from the standard curve. Calculate results as μg. cadmium/ml. of leaching solution.

A sample is considered violative if the average of the 3 units examined contains 0.5 μg. cadmium/ml. of leaching solution or more.

APPENDIX A–7

THE UNITED STATES POTTERS ASSOCIATION CERAMIC DINNERWARE SURVEILLANCE PROGRAM*

I. Test Procedure

The United States Food and Drug Administration Method LIB-834 shall be the procedure for determining the release of heavy metals from ceramic surfaces. *Exception:* Instead of 6 units individually analyzed, as indicated in FDA-LIB-834, the leachate from 3 units shall be blended for the determination of heavy metal release.

II. Tolerance Standards

A. The maximum allowable lead release from ceramic surfaces for use in food and beverage service shall be less than 7 ppm of the leaching solution when determined by FDA-LIB-834.

B. A safety zone for lead release shall be established between 6 and 7 ppm. A finding of lead release in this range shall be cause to alert the producer to reformulate or to alter the firing procedure, or both, in order to reduce the lead release to a value below 6 ppm.

C. The maximum allowable cadmium release from ceramic surfaces for use in food and beverage service shall be less than 0.5 ppm when determined by FDA-LIB-834.

III. Surveillance Procedures

A. Sampling and Testing

1. a. Members of The United States Potters Association Ceramic Dinnerware Surveillance Program (USPA) shall twice each year submit three representative current production samples of each different type of ceramic surface designed for use in food and beverage service. Responsibility for periodically submitting such test specimens shall rest upon each mem-

*Courtesy Lead Industries Association, Inc., 292 Madison Avenue, New York, N.Y.

ber of the USPA Ceramic Dinnerware Surveillance Program. Testing shall be done by the Pittsburgh Testing Laboratory, 850 Poplar St., Pittsburgh, Pa. 15220; The Twinning Laboratories, Inc., Box 1472, Fresno, California, 93716; or other laboratories approved by the USPA. Samples shall be submitted before February 1 and August 1, with reporting of the results by February 15 and August 15.

b. All new ceramic surfaces for use in food and beverage service made by members of the USPA. Ceramic Dinnerware Surveillance Program shall be tested and found to meet USPA standards for heavy metal release before they may be placed in production.

c. A copy of the submittal letter or order relating to each sample submitted for test shall be sent to the Chairman of the USPA Research Committee.

d. A copy of the results of each test conducted in the Ceramic Dinnerware Surveillance Program shall be sent to the Chairman of the USPA Research Committee by the testing laboratory.

e. Cost of tests on participants own ware shall be borne by the individual members of the USPA Ceramic Dinnerware Surveillance Program.

f. Members of the USPA Ceramic Dinnerware Surveillance Program shall be responsible for sampling suspected ceramic ware made by nonparticipating suppliers, both foreign and domestic.

   The USPA shall bear the cost of both samples and testing of suspect glazed ware by members of the USPA Ceramic Dinnerware Surveillance Program. Costs incurred by nonmembers of the USPA shall be borne by the individual members of any Association, apart from the USPA Ceramic Dinnerware Surveillance Program, they may be connected with. USPA members shall first inform the Chairman of the USPA Research Committee of suspect ware and with his permission then submit samples of the suspect ware for test of heavy metal release.

2. The USPA shall endeavor to enlist all suppliers of ceramic

dinnerware to the American market as members of the USPA Ceramic Dinnerware Surveillance Program. All shall participate on an equal basis whether the ware supplied is of domestic or foreign production.

B. Approval

1. When a ceramic surface for use in food and beverage service is found to show lead release of 6 ppm of the leaching solution or less and cadmium release less than 0.5 ppm, the USPA shall license the use of a suitable mark signifying that the ware has been tested and approved by the USPA. This mark may be a backstamp or label to be applied to the ware or may be a mark to be imprinted on the shipping containers or in advertising material.
2. The USPA mark of acceptance shall be available to all whose ware meets the USPA standards of heavy metal release and shall relate specifically to each individual type of ceramic surface.

C. Policing

1. Whenever a test report discloses lead release between 6 and 7 ppm from a ceramic surface designed for food or beverage service and produced by a member of the USPA Ceramic Dinnerware Surveillance Program, the maker shall be requested to take prompt remedial action.
2. Whenever a test report discloses a lead release of 7 or more ppm from ceramic surface designed for food or beverage service and produced by a member of the USPA Ceramic Dinnerware Surveillance Program, the producer and the Food and Drug Administration, Office of Compliance Programs, shall be advised promptly of the fact so that appropriate action may be taken.
3. Whenever a test report on suspect ware which may be used for food or beverage service and produced by a supplier not a member of the USPA Ceramic Dinnerware Surveillance Program discloses lead release of 7 or more ppm, the Chairman of the USPA Research Committee shall promptly notify the FDA Office of Compliance Programs and the producer.

4. Whenever a test report discloses a cadmium release of 0.5 or more from a ceramic surface designed for food or beverage service and produced by a member of the USPA Ceramic Dinnerware Surveillance Program, the producer and the FDA Office of Compliance Programs shall be advised promptly of the fact so that appropriate action may be taken.
5. Whenever a test report on suspect ware which may be used in food or beverage service and produced by a supplier not a member of the USPA Ceramic Dinnerware Surveillance Program discloses cadmium release of 0.5 or more ppm, the Chairman of the USPA Research Committee shall promptly notify the FDA Office of Compliance Programs and the producer.
6. When it is evident that any ceramic surface tested contains hazardous metals other than lead and/or cadmium, the quantity released under the test conditions of FDA-LIB-834 shall also be determined and reported.
7. Should a member of the USPA Ceramic Dinnerware Surveillance Program, but not a member of USPA, fail to submit the annual license fee, such member's license to use the USPA mark of acceptance shall be terminated.
8. Whenever a member of the USPA Ceramic Dinnerware Surveillance Program has been notified under C-1, C-2 or C-4 to correct an excessive lead or cadmium release from a ceramic surface, failure by such member to promptly effect a reduction to prescribed limits, shall be sufficient cause to terminate such member's privilege to use the USPA mark of acceptance for that ware.

IV. Reporting to FDA

1. The Chairman of the USPA Research Committee shall issue a report to the FDA twice each year. These reports shall give an account of its activities, showing tests performed, the results and actions taken by the USPA or proposed to the FDA. The reports shall be issued by March 15 and September 15 each year.
2. In addition to the regular semi-annual reports, the Chairman of the USPA Research Committee shall report promptly to the

FDA Office of Compliance Programs any test results showing 7 or more parts of lead release per million parts of the leaching solution and/or 0.5 or more parts of cadmium release per million parts of the leaching solution. Any significant quantities of other heavy metals shall also be reported. This report shall include a statement of any action which the USPA may propose to the FDA in the event the USPA is powerless to take effective action.

V. Revisory Provisions

1. Stipulations set forth in this program may be amended by only the USPA when warranted. Members of the USPA Ceramic Dinnerware Surveillance Program shall be promptly advised of such changes.

May 20, 1971

APPENDIX A–8

CONE 06 ACID RESISTANT GLAZES

Cone 06 glazes have consistently demonstrated lead release of less than 7 ppm even when fired in a static atmosphere such as in an electric kiln. Any other low temperature glaze formulations should be tested before being considered safe for food and beverage applications.*

CONE 06 ACID RESISTANT GLAZES
(GLAZES MADE FROM 90% FRIT AND 10% KAOLIN)

Molecular Formula of Glaze

	Na_2O	CaO	SrO	PbO	Al_2O_3	B_2O_3	SiO_2	ZrO_2
1	0.05	0.06	—	0.89	0.27	0.15	2.58	0.01
2	0.10	0.14	—	0.76	0.27	0.32	2.69	0.02
3	0.05	0.07	0.22	0.72	0.27	0.24	2.68	0.01
4	0.16	0.21	—	0.63	0.28	0.49	2.80	0.03
5	0.10	0.13	0.14	0.62	0.27	0.36	2.65	0.02
6	0.16	0.20	0.08	0.56	0.27	0.51	2.78	0.03

Weight Percent Formula of Glazes

	Na_2O	CaO	SrO	PbO	Al_2O_3	B_2O_3	SiO_2	ZrO_2
1	0.78	0.93	—	49.53	6.96	2.71	38.83	0.25
2	1.61	1.92	—	42.69	7.03	5.55	40.67	0.52
3	0.81	0.96	5.78	40.51	7.11	4.22	40.34	0.26
4	2.49	2.94	—	35.51	7.11	8.54	42.61	0.80
5	1.64	1.96	3.89	36.47	7.13	6.63	41.74	0.53
6	2.51	2.97	1.97	32.31	7.14	9.12	43.17	0.81

Note: While all of these glazes are clear, numbers 1, 2 and 3 develop a slight yellowish cast. Formulations 4, 5 and 6 are clear and colorless.

*Courtesy Lead Industries Association, Inc., 292 Madison Ave., New York, N.Y.

APPENDIX A–9

TEMPERATURE CONVERSIONS†

C	*	F	C	*	F	C	*	F	C	*	F
−273.15	−459.67		−118	−180	−292	−11.7	11	51.8	4.4	40	104.0
−268	−450		−112	−170	−274	−11.1	12	53.6	5.0	41	105.8
−262	−440		−107	−160	−256	−10.6	13	55.4	5.6	42	107.6
−257	−430		−101	−150	−238	−10.0	14	57.2	6.1	43	109.4
−251	−420		−95.6	−140	−220	−9.4	15	59.0	6.7	44	111.2
−246	−410		−90.0	−130	−202	−8.9	16	60.8	7.2	45	113.0
−240	−400		−84.4	−120	−184	−8.3	17	62.6	7.8	46	114.8
−234	−390		−78.9	−110	−166	−7.8	18	64.4	8.3	47	116.6
−229	−380		−73.3	−100	−148	−7.2	19	66.2	8.9	48	118.4
−223	−370		−67.8	−90	−130	−6.7	20	68.0	9.4	49	120.2
−218	−360		−62.2	−80	−112	−6.1	21	69.8	10.0	50	122.0
−212	−350		−56.7	−70	−94	−5.6	22	71.6	10.6	51	123.8
−207	−340		−51.1	−60	−76	−5.0	23	73.4	11.1	52	125.6
−201	−330		−45.6	−50	−58	−4.4	24	75.2	11.7	53	127.4
−196	−320		−40.0	−40	−40	−3.9	25	77.0	12.2	54	129.2
−190	−310		−34.4	−30	−22	−3.3	26	78.8	12.8	55	131.0
−184	−300		−28.9	−20	−4	−2.8	27	80.6	13.3	56	132.8
−179	−290		−23.3	−10	14	−2.2	28	82.4	13.9	57	134.6
−173	−280		−17.8	0	32	−1.7	29	84.2	14.4	58	136.4
−169	−273	−459.4	−17.2	1	33.8	−1.1	30	86.0	15.0	59	138.2
−168	−270	−454	−16.7	2	35.6	−0.6	31	87.8	15.6	60	140.0
−162	−260	−436	−16.1	3	37.4	0	32	89.6	16.1	61	141.8
−157	−250	−418	−15.6	4	39.2	0.6	33	91.4	16.7	62	143.6
−151	−240	−400	−15.0	5	41.0	1.1	34	93.2	17.2	63	145.4
−146	−230	−382	−14.4	6	42.8	1.7	35	95.0	17.8	64	147.2
−140	−220	−364	−13.9	7	44.6	2.2	36	96.8	18.3	65	149.0
−134	−210	−346	−13.3	8	46.4	2.8	37	98.6	18.9	66	150.8
−129	−200	−328	−12.8	9	48.2	3.3	38	100.4	19.4	67	152.6
−123	−190	−310	−12.2	10	50.0	3.9	39	102.2	20.0	68	154.4

INTERPOLATION VALUES

C	*	F	C	*	F
0.56	1	1.8	3.33	6	10.8
1.11	2	3.6	3.89	7	12.6
1.67	3	5.4	4.44	8	14.4
2.22	4	7.2	5.00	9	16.2
2.78	5	9.0	5.56	10	18.0

*In the center column, find the temperature to be converted. The equivalent temperature is in the left column, if converting to Celsius, and in the right column, if converting to Fahrenheit.

†From *Instrumentation for Process Measurement and Control*, Third Edition, by Norman A. Anderson. Copyright 1980 by the author. Reprinted with the permission of the publisher, Chilton Book Company, Radnor, Pennsylvania.

APPENDIX A–9
TEMPERATURE CONVERSIONS (*continued*)

C	★	F	C	★	F	C	★	F	C	★	F
20.6	69	156.2	127	260	500	443	830	1526	760	1400	2552
21.1	70	158.0	132	270	518	449	840	1544	766	1410	2570
21.7	71	159.8	138	280	536	454	850	1562	771	1420	2588
22.2	72	161.6	143	290	554	460	860	1580	777	1430	2606
22.8	73	163.4	149	300	572	466	870	1598	782	1440	2624
23.3	74	165.2	154	310	590	471	880	1616	788	1450	2642
23.9	75	167.0	160	320	608	477	890	1634	793	1460	2660
24.4	76	168.8	166	330	626	482	900	1652	799	1470	2678
25.0	77	170.6	171	340	644	488	910	1670	804	1480	2696
25.6	78	172.4	177	350	662	493	920	1688	810	1490	2714
26.1	79	174.2	182	360	680	499	930	1706	816	1500	2732
26.7	80	176.0	188	370	698	504	940	1724	821	1510	2750
27.2	81	177.8	193	380	716	510	950	1742	827	1520	2768
27.8	82	179.6	199	390	734	516	960	1760	832	1530	2786
28.3	83	181.4	204	400	752	521	970	1778	838	1540	2804
28.9	84	183.2	210	410	770	527	980	1796	843	1550	2822
29.4	85	185.0	216	420	788	532	990	1814	849	1560	2840
30.0	86	186.8	221	430	806	538	1000	1832	854	1570	2858
30.6	87	188.6	227	440	824	543	1010	1850	860	1580	2876
31.1	88	190.4	232	450	842	549	1020	1868	866	1590	2894
31.7	89	192.2	238	460	860	554	1030	1886	871	1600	2912
32.2	90	194.0	243	470	878	560	1040	1904	877	1610	2930
32.8	91	195.8	249	480	896	566	1050	1922	882	1620	2948
33.3	92	197.6	254	490	914	571	1060	1940	888	1630	2966
33.9	93	199.4	260	500	932	577	1070	1958	893	1640	2984
34.4	94	201.2	266	510	950	582	1080	1976	899	1650	3002
35.0	95	203.0	271	520	968	588	1090	1994	904	1660	3020
35.6	96	204.8	277	530	986	593	1100	2012	910	1670	3038
36.1	97	206.6	282	540	1004	599	1110	2030	916	1680	3056
36.7	98	208.4	288	550	1022	604	1120	2048	921	1690	3074
37.2	99	210.2	293	560	1040	610	1130	2066	927	1700	3092
37.8	100	212.0	299	570	1058	616	1140	2084	932	1710	3110
43	110	230	304	580	1076	621	1150	2102	938	1720	3128
49	120	248	310	590	1094	627	1160	2120	943	1730	3146
54	130	266	316	600	1112	632	1170	2138	949	1740	3164
60	140	284	321	610	1130	638	1180	2156	954	1750	3182
66	150	302	327	620	1148	643	1190	2174	960	1760	3200
71	160	320	332	630	1166	649	1200	2192	966	1770	3218
77	170	338	338	640	1184	654	1210	2210	971	1780	3236
82	180	356	343	650	1202	660	1220	2228	977	1790	3254
88	190	374	349	660	1220	666	1230	2246	982	1800	3272
93	200	392	354	670	1238	671	1240	2264	988	1810	3290
99	210	410	360	680	1256	677	1250	2282	993	1820	3308
			366	690	1274	682	1260	2300	999	1830	3326
			371	700	1292	688	1270	2318	1004	1840	3344
			377	710	1310	693	1280	2336	1010	1850	3362
			382	720	1328	699	1290	2354	1016	1860	3380
100	212	413	388	730	1346	704	1300	2372	1021	1870	3398
			393	740	1364	710	1310	2390	1027	1880	3416
			399	750	1382	716	1320	2408	1032	1890	3434
			404	760	1400	721	1330	2426	1038	1900	3452
			410	770	1418	727	1340	2444	1043	1910	3470
			416	780	1436	732	1350	2462	1049	1920	3488
104	220	428	421	790	1454	738	1360	2480	1054	1930	3506
110	230	446	427	800	1472	743	1370	2498	1060	1940	3524
116	240	464	432	810	1490	749	1380	2516	1066	1950	3542
121	250	482	438	820	1508	754	1390	2534	1071	1960	3560

C	*	F	C	*	F	C	*	F	C	*	F
1077	1970	3578	1221	2230	4046	1366	2490	4514	1510	2750	4982
1082	1980	3596	1227	2240	4064	1371	2500	4532	1516	2760	5000
1088	1990	3614	1232	2250	4082	1377	2510	4550	1521	2770	5018
1093	2000	3632	1238	2260	4100	1382	2520	4568	1527	2780	5036
1099	2010	3650	1243	2270	4118	1388	2530	4586	1532	2790	5054
1104	2020	3668	1249	2280	4136	1393	2540	4604	1538	2800	5072
1110	2030	3686	1254	2290	4154	1399	2550	4622	1543	2810	5090
1116	2040	3704	1260	2300	4172	1404	2560	4640	1549	2820	5108
1121	2050	3722	1266	2310	4190	1410	2570	4658	1554	2830	5126
1127	2060	3740	1271	2320	4208	1416	2580	4676	1560	2840	5144
1132	2070	3758	1277	2330	4226	1421	2590	4694	1566	2850	5162
1138	2080	3776	1282	2340	4244	1427	2600	4712	1571	2860	5180
1143	2090	3794	1288	2350	4262	1432	2610	4730	1577	2870	5198
1149	2100	3812	1293	2360	4280	1438	2620	4748	1582	2880	5216
1154	2110	3830	1299	2370	4298	1443	2630	4766	1588	2890	5234
1160	2120	3848	1304	2380	4316	1449	2640	4784	1593	2900	5252
1166	2130	3866	1310	2390	4334	1454	2650	4802	1599	2910	5270
1171	2140	3884	1316	2400	4352	1460	2660	4820	1604	2920	5288
1177	2150	3902	1321	2410	4370	1466	2670	4838	1610	2930	5306
1182	2160	3920	1327	2420	4388	1471	2680	4856	1616	2940	5324
1188	2170	3938	1332	2430	4406	1477	2690	4874	1621	2950	5342
1193	2180	3956	1338	2440	4424	1482	2700	4892	1627	2960	5360
1199	2190	3974	1343	2450	4442	1488	2710	4910	1632	2970	5378
1204	2200	3992	1349	2460	4460	1493	2720	4928	1638	2980	5396
1210	2210	4010	1354	2470	4478	1499	2730	4946	1643	2990	5414
1216	2220	4028	1360	2480	4496	1504	2740	4964	1649	3000	5432

TEMPERATURE CONVERSION FORMULAS

Degrees Celsius (Formerly Centigrade) C	**Degrees Fahrenheit—F**
$(C \times 9/5) + 32 = F$ Fahrenheit	$(F - 32) \times 5/9 = C$ Celsius

APPENDIX A–10
GLOSSARY

Albite A soda feldspar having the composition $Na_2O \cdot Al_2O_3 \cdot 6SiO_2$.

Amorphous A state of matter that has no regular repetitive arrangement of atoms or molecules; noncrystalline matter, such as glass or liquids.

Associated Liquid One in which the molecules have an affinity or attraction for one another resulting in the build-up of some orderly arrangement over short distances.

Auger A device that forces a material through an opening in a die; usually a rotating screw implement.

Ball Clay A fine grained sedimentary clay characterized by a high degree of plasticity and dry strength and a high drying shrinkage.

Base Exchange The property of a material to adsorb ions from the surrounding solution. Kaolin and other clay minerals have this property, and the amount of ions adsorbed on their surfaces is called the base exchange capacity.

Beneficiation A process whereby unwanted impurities are removed from a material.

Black Core Insufficient oxygen in the kiln atmosphere during the early stages of firing (450–600°C) causes leucoxene to change to suboxides of iron and titanium, which are gray, blue or black; thus, the terms gray coring, blue coring or black coring. Oxygen penetrates the ware from the outside, so the centers of thick ware sections may be discolored after firing because of insufficient air or time in the range 450 to 600°C. Normally leucoxene oxidizes and causes ware to be cream-colored.

Blue Core See Black Core.

Blunging The process of mixing a material, such as clay, with a liquid, such as water. Usually done with propeller-type or other high speed mixers.

Calcine To treat a material before use by heating; for example, by calcining a clay, the water is removed, shrinkage has taken place and the material undergoes no further change until heated above its original calcining temperature.

Calcium Hydroxide See Lime.

Chlorite A class of platy minerals similar to mica in appearance but much softer and of quite different compositions. Chlorites are hydrated silicates of magnesium and aluminum with a proportion of iron in either the ferrous or ferric form. Crystals are of small grain size and are greenish in color.

Combustion The process in which a mixture of oxygen and organic material combines and burns thereby giving off heat.

Conduction The transfer of heat through a material. The method by which the interior of a piece becomes heated.

Convection The transfer of heat due to the flow of hot gases over or past a cooler piece of material. The major means of heat transfer below red heat.

Cordierite A crystalline material having the composition $2MgO \cdot 2Al_2O_3 \cdot 5SiO_2$ and having a low thermal expansion. Cordierite is formed when talc and clay are fired together.

Crawling A type of glaze defect in which the glaze does not wet or adhere to the surface of the body resulting in unglazed areas of body.

Crazing A glaze defect resulting from the glaze being in too much tension which causes a fine "hairline" crack pattern in the glaze.

Cristobalite A high-temperature form of quartz, SiO_2, which results when quartz is heated in the 1100 to 1200°C range. This inversion is aided by the presence of CaO or Fe_2O_3. This is a sluggish, nonreversible inversion. Cristobalite has a sharp thermal expansion discontinuity in the 200 to 300°C range, which is responsible for thermal shock problems in many clay bodies.

Crystal A form of matter having a characteristic shape with the atoms having a repeating symmetrical arrangement throughout the mass.

Deflocculation The causing of particles to repel each other and, hence, remain as far apart as possible. The addition of certain chemicals that cause particles to develop similar charges causes deflocculation. This results in stable suspensions that do not settle.

Dehydroxylation Removal of the OH^- ions from the crystal structure of minerals. Dehydroxylation of kaolinite occurs in the 450 to 600°C range.

Differential Water Variation in water content in different areas of the same piece.

Dipole An object having opposite charges on the two extremities, for example, the north and south poles of a magnet, or the water molecule, which exhibits a positive charge on one side and a negative charge on the other. Several organic molecules show this property.

Drier Scum A white deposit that forms on the ware during the drying process. It is due to soluble salts in the body that are brought to the surface along with the water during drying.

Dunting A crack that occurs while ware is cooling, or bisque ware is reheating, in the 400 to 600°C range. It is caused by the nonlinear expansion and contraction of quartz in this temperature range.

Endothermic A reaction that requires heat in order to take place, such as the dehydroxylation of kaolin.

$$Al_2O_3{\cdot}2SiO_2{\cdot}2H_2O + \text{Heat} \rightarrow Al_2O_3{\cdot}2SiO_2 + 2H_2O$$

Engobe A coating applied to the body before glazing; engobes usually consist of china clay, ball clay, feldspar or whiting.

Excess Air Air above that required for complete combustion of fuel. Air required to maintain oxidizing conditions in the kiln atmosphere.

Exchange Capacity A measure of the ability of a particle, such as clay, to adsorb on its surface ions present in the surrounding medium. In the case of clay minerals this is primarily a function of the particle size or surface area of the particle.

Exfoliation The property exhibited by some layer minerals, such as vermiculite, which, when heated, expand or swell perpendicular to the plates like small accordions.

Exothermic A chemical reaction that evolves heat, such as the oxidation of carbon (burning of wood or coal). $C + O_2 \rightarrow CO_2 + \text{Heat}$

Expansion Coefficient The amount that 1 in. of a material will increase in length when heated 1°F.

Extensibility The ability of a material to be deformed without forming planes of weakness or cracks.

Extrusion The process of forcing a plastic material through a die or orifice resulting in some predetermined shape.

Fire Clay A clay that varies widely in composition, resulting in three main types: flint, plastic and high alumina. *Flint* fire clay is hard and shows little plasticity when finely ground. When mixed with plastic clay, it serves well as a grog to reduce shrinkage. *Plastic* fire clays have wide variations in composition resulting in a wide range in fusion points. Thus, there are "low" or "high" heat duty fire clays depending upon the amount of fluxing impurities they contain. *High Alumina* fire clays, because of their ability to withstand high temperatures, are used in the manufacture of super-duty alumina brick.

Flocculation The causing of particles to come together and form flocs. This results when the particles have no charge, do not repel each other, form flocs composed of many particles and settle out rapidly.

Flux A material that reduces the melting point of the material to which it is added. For example, the addition of Na_2O to SiO_2 reduces the melting point.

Free Water The water added to a system that is removed by evaporation.

Frit A frit is a partially fused mixture of two or more materials, some or all of which could not be used separately on account of their solubility in water. In the process of fritting, chemical combination occurs and insoluble substances are formed.

Gray Core See Black Core.

Hydroxyl The $(OH)^-$ ion is the hydroxyl ion. Two such ions can combine to form a water molecule. Hence, dehydroxylation of a clay means simply removal of the $(OH)^-$ ions from the crystal structure by heating in the 450 to 600°C range.

Hygroscopic The property that causes a material to readily pick up moisture.

Illite A mica-like mineral with less alkali ions than muscovite, of small particle size and having a larger water content than true micas. Sometimes called hydrous mica.

Kaolinite One of the principal clay minerals having the composition $Al_2O_3 \cdot 2SiO_2 \cdot 2H_2O$.

Kiln Atmosphere The gaseous environment surrounding the ware in a kiln.

Laminations Discontinuities in a system resulting in planes of weakness.

Leucite An intermediate phase or crystal that forms when potash feldspar is heated to 1150°C. This crystal does not completely melt until 1520°C is reached. It has the composition $K_2O \cdot Al_2O_3 \cdot 4SiO_2$.

Leucoxene A black, amorphous, gelatinous material found between layers in kaolinite plates. It contains suboxides of iron and titanium and is responsible for the darkening of the fired color in U.S. clays.

Lime Calcium oxide, CaO, sometimes called quick lime. When water is added it reacts rapidly to form calcium hydroxide, Ca(OH).

Lime Popping A surface defect resulting from the rehydration of a lime (CaO) particle to form calcium hydroxide ($Ca(OH)_2$). Rehydration causes a large expansion of the lime particle, producing enough stress to rupture the area surrounding it.

Memory The property of a plastic clay-water mix that causes it to attempt to return to the shape it had prior to its most recent deformation.

Metakaolin Dehydroxylated kaolin. It is nonplastic, has lost its crystallinity but retains its shape and will readsorb water to again form kaolin.

Methane A gas having the composition CH_4; the main constituent of natural gas.

Mica A platelike mineral that easily cleaves into thin sheets and has the composition $K_2O \cdot 3Al_2O_3 \cdot 6SiO_2 \cdot 2H_2O$ (muscovite mica).

Microcline A potash feldspar having the composition $K_2O \cdot Al_2O_3 \cdot 6SiO_2$.

Micron (μ) A micron is 0.001 mm. or 0.00001 cm.

Modulus of Elasticity The measure of the capacity of a body to withstand stress without deformation. Ceramic bodies can withstand high stress without deformation and, therefore, have a high modulus of elasticity.

Mole A mole is one molecular weight of a material expressed in grams. One mole of Al_2O_3 is 102 g.

Montmorillonite Commonly called bentonite, characterized by its extremely small particle size and having the composition $Al_2O_3 \cdot 4SiO_2 \cdot H_2O$.

Muscovite See Mica.

Nepheline Syenite A raw mineral mixture of approximately 50 percent albite, 25 percent microcline, 25 percent nepheline and no quartz. It has the composition $Na_2O \cdot Al_2O_3 \cdot 2SiO_2$.

Opacifier A material that reduces light transmission through the host material. In glazes, opacifiers are used to cover ware blemishes, intensify colors and produce special effects.

Orientation The process of causing the orderly arrangement of particles.

Orthoclase A potash feldspar having the composition $K_2O \cdot Al_2O_3 \cdot 6SiO_2$.

Perfect Combustion The burning of a mixture of air and fuel in the proper ratio to attain complete combustion.

Petalite A lithium-alumina-silicate mineral having the composition $Li_2O \cdot Al_2O_3 \cdot 8SiO_2$.

pH The measure of acid-base balance of a liquid on a scale of 1 to 14, with 1 to 6 being acid, 7 neutral and 8 to 14 base.

Plasticity The property of a material that allows deformation and retention of the deformed shape.

Plucking A defect in the ware caused by the fusion and adherence of the ware to the surface upon which it has been fired.

Poisson's Ratio The ratio between the decrease in width and the increase in length of a cross section of a bar as it undergoes tensile stress. An indication of how much a material will neck down when pulled in tension. Because ceramic materials flow very little, their Poisson's ratio is small.

Pore Water Water that remains in a piece after shrinkage is complete and the particles are in contact with each other.

Porosity Pores or voids in the ware resulting from interconnected channels.

Potter's Flint Finely ground silica, SiO_2.

Primary Air Air mixed with fuel at the burner.

Primary Mullite Mullite that crystallizes from the kaolinite particles.

Radiation The transfer of heat through space by infrared rays. The main method of heat transfer above red heat, approximately 600°C.

Refractoriness The ability of a material to withstand high temperature.

Rehydration The process of regaining water that had originally been removed from a material by heating.

Salt Glaze A coating developed on a clay body caused by the reaction of sodium chloride vapor with the alumina-silica body, resulting in a soda-alumina-silicate glass coating.

Secondary Air Air admitted to the kiln through openings in the kiln walls.

Secondary Mullite Mullite that crystallizes from the alumina and silica present in the glassy phase.

Segregation The tendency of materials to separate due to differences in density or particle size.

Set Point The temperature at which the viscosity of a glass being cooled becomes so high that it behaves as a rigid solid. Sometimes called the strain point.

Shivering A glaze defect caused by too much compression in the glaze.

Shrinkage Water The water between particles in a clay-water system which when removed by drying causes the particles to come closer together, thus causing shrinkage. The volume percent shrinkage equals the volume of shrinkage water removed during drying.

Silica (Glassy) When heated to a sufficiently high temperature, quartz melts to a viscous liquid. When cooled, this liquid does not crystallize and silica glass is formed.

Slumping Deformation of a material caused by overfiring.

Solubility Product A number that defines the limits of solubility of the ions in a slightly soluble material, i.e., ${C_{Ca}}^{+2} \times C_{SO_4=} = 1.95 \times 10^{-4}$ where (C) represents the molar concentration of the ions.

Solvated Layer A water film immediately surrounding a particle surface.

Specific Gravity The weight of a given volume of a material compared to an equal volume of water. One cc. of water weighs 1 g.; thus, the specific gravity of water is 1.

Spinel A type of oxide crystal structure, AB_2O_4, for example, $MgAl_2O_4$.

Spodumene A lithium-alumina-silicate mineral having the composition $Li_2O \cdot Al_2O_3 \cdot 4SiO_2$. In combination with clay, it is most useful in the fabrication of low-expansion ceramic bodies.

Surface Tension The property observable in liquids which results in the surface acting as though it were in tension like a stretched membrane. The reason is that the molecules on the surface are not symmetrically surrounded and are attracted only in the direction of the interior of the liquid.

Tensile Strength The force in pounds required to break a unit cross section of a piece in tension; units lbs./in.^{2}.

Terra Sigillata A type of coating prepared from the extremely fine particles of a clay.

Thermal Conductivity A measure of the ability of a material to transfer heat; Units BTU $hr.^{-1}ft.^{2}F^{\circ -1}ft.$

Thermal Expansion That property of a material to expand when heated and shrink when cooled.

Thermogravimetric The effect of heat on the weight loss of a material.

Viscosity The resistance shown by a fluid to flow or movement. The higher the viscosity the more resistance to flow.

Vitrification The formation of a glassy phase as a result of heating a material.

Water of Plasticity The water required by a material to develop maximum plasticity.

Wedging The process of repeated cutting, separating and reuniting plastic clay for the purpose of providing completely random orientation of particles and uniform distribution of moisture.

Yield Point The force required to start deformation in a system.

INDEX

W. G. Lawrence

Willis Grant Lawrence, a member and Fellow of the American Ceramic Society and the American Foundrymen's Society, graduated from the New York State College of Ceramics at Alfred University and received the Sc.D. degree from the Massachusetts Institute of Technology. Dr. Lawrence has been active in research, teaching and administration at the College of Ceramics at Alfred University for the past 33 years. He is the author of one other book and numerous papers on subjects related to clay-water systems, building materials and mold materials for the precision casting of metals. Early in his career as a ceramic engineer with the Abex Corporation, the author was appointed Professor of Research at the College of Ceramics in 1947. In June 1980 he retired as Dean of the College of Ceramics and now holds the title of Dean Emeritus.

In 1972 he received the coveted Jeppson Award of the American Ceramic Society for his contributions to the field of ceramic education and science.

Richard R. West

Richard R. West, Professor Emeritus, New York State College of Ceramics at Alfred University, taught, conducted research and consulted in the field of ceramic whiteware for 35 years. He is a licensed professional engineer, a Fellow of the American Ceramic Society and has written over 30 technical papers as well as contributed chapters for a number of monographs. He has specialized in the field of testing and analyzing clays and for 10 years was the chairman of the committee on testing ceramic whiteware clays for the American Society for Testing and Materials.